Columbia University

Contributions to Education

Teachers College Series

No. 117

AMS PRESS
NEW YORK

The Place of the Elementary Calculus in the Senior High-School Mathematics

And Suggestions for a Modern Presentation of the Subject

By

NOAH BRYAN ROSENBERGER, Ph.D.

Teachers College, Columbia University
Contributions to Education, No. 117

Published by
Teachers College, Columbia University
New York City
1921

Library of Congress Cataloging in Publication Data

Rosenberger, Noah Bryan, 1881-
The place of the elementary calculus in the senior high-school mathematics and suggestions for a modern presentation of the subject.

Reprint of the 1921 ed., issued in series: Teachers College, Columbia University. Contributions to education, no. 117.
Originally presented as the author's thesis, Columbia.
Includes bibliographical references.
1. Calculus--Study and teaching (Secondary) I. Title. II. Series: Columbia University. Teachers College. Contributions to education, no. 117.

QA303.R7 1972 515'.07'12 71-177209
ISBN 0-404-55117-3

Reprinted by Special Arrangement with Teachers College Press, New York, New York

From the edition of 1921, New York
First AMS edition published in 1972
Manufactured in the United States

AMS PRESS, INC.
NEW YORK, N. Y. 10003

PREFACE

EVEN a little less than a decade ago, the trained teacher of high-school mathematics had the pleasure of teaching real mathematics as found in such subjects as trigonometry, solid and plane geometry, and advanced and elementary algebra. It is doubtless true that this kind of work was inclining too much toward formalism and that the needs of the learner were not being given sufficient attention. The strong reaction against formalism in general in education during the decade just past made the position of any genuine mathematics in our high schools very precarious. This threatened elimination of all real mathematics in our high schools was viewed with regret by those of us who had had the pleasure of starting many a pupil with a vigorous mind upon his mathematical way; and a number of prominent educators who are especially interested in the teaching of mathematics have been working toward a readjustment of the mathematics curriculum so that it will fit into and fulfill its mission in the changed conditions as found in our schools to-day.

The following study is the author's contribution toward this attempted readjustment. The kind of mathematics that should be taught in the first six school years has been definitely settled; the main points of the work in mathematics that should be included in the junior high-school period have been agreed upon; and the author hopes that his study will be of some value in helping us to formulate the content of the mathematics curriculum of the senior high school.

The author takes pleasure in acknowledging his indebtedness to Professor David Eugene Smith for his wise and constructive criticism and helpful guidance in the preparation of this study.

NOAH BRYAN ROSENBERGER
Acting Head of Department of Mathematics,
Dickinson College, Carlisle, Pa.

July 1, 1921

TABLE OF CONTENTS

INTRODUCTION

A STUDY of the nature of the courses in mathematics found from time to time in our high schools will show marked changes in the kind and extent of the topics included in these courses. These changes will be indicated in a general manner in the discussion which follows. In colonial times, in the schools corresponding approximately to our high schools, the arithmetic consisted of operations with integral numbers and of some work in fractions and in proportion. A few decades later, when our high schools first began to make their appearance, arithmetic had become more difficult and included such topics as stocks and exchange, circulating decimals, alligation, cube root, and permutations and combinations. Then it was realized that it would be better not to teach so much of the difficult parts of arithmetic but more of the simpler parts of some of the higher branches in mathematics. Hence, during the latter part of the nineteenth century, algebra and geometry became a part of the high-school mathematics.[1] These were followed a number of years later by the introduction of trigonometry.

In our present decade there is taking place an extended reorganization in our school curricula. One of the phases of this reorganization, namely, the junior high-school movement, has caused a radical change in the content of the mathematics curriculum. It has brought into the courses in mathematics in the seventh, eighth, and ninth school years the simpler parts of geometry, algebra, and trigonometry. This makes necessary an entire rearrangement of the courses in mathematics in the tenth, eleventh, and twelfth school years, or the senior high-school period. Such subjects as demonstrative geometry, algebra, and trigonometry are assured of a place in the mathematics of the senior high school. However, these subjects will not now be sufficient to meet the needs of pupils who wish to elect courses in mathematics in each of these three years. These needs will probably be met, as similar ones have been in the past, by the introduction of the simpler parts of some of the higher branches

[1] Paul Monroe (ed.), *A Cyclopedia of Education*, Vol. 3, pp. 263 and 267, New York, 1912; Florian Cajori, *The Teaching and History of Mathematics in the United States*, p. 9, Washington, 1890.

in mathematics; and it is the aim of this study to show that among these should be included the simpler parts of the calculus.

GENERAL STEPS IN THE ARGUMENT

I. The first topic will be a discussion based upon a study of the status of mathematics in foreign schools corresponding to our high schools. This will show that in many countries the calculus is taught in these schools before the end of the year which corresponds to our twelfth school year. This study will also give us an outside standard whereby to judge the status of mathematics in our schools.

II. The second topic will be a discussion of the trend of mathematics in our public school system. This will show that there has been created an opportunity for including the elementary calculus in the senior high-school mathematics. The question then arises, since there is an opportunity for doing this, is the subject of sufficient importance?

III. The third topic will be a discussion of the important position occupied by the calculus in the mathematics structure. This will show that the calculus is one of the most important aids in applied mathematics and contributes largely to the field of pure mathematics. The conclusion will be drawn that, because of its importance, senior high-school pupils who are interested in mathematics should be given an opportunity to become acquainted with the elementary calculus, provided it is not too difficult.

IV. The fourth topic will give a historical survey of the natural growth of the calculus in the development of mathematics. This will show that a first course in the elementary calculus is well within the mental grasp of pupils in mathematics in the senior high school. This historical survey will also suggest suitable methods of teaching the elementary calculus to younger pupils.

V. The fifth topic will give some of the features of various textbooks on the elementary calculus that have been written for beginners. It will show that a number of professors and able teachers of mathematics do not consider the elementary calculus too difficult for younger pupils. This comparison of textbooks will also be of value in the selection of methods of presentation of the elementary calculus.

VI. The sixth topic will indicate the trend of American education in general. It will show that among many people educational activity does not cease at the close of their public school life. Of course these later mental pursuits follow many different channels. A person working in the electrical engineering field will be likely to take up studies along the lines of electricity and engineering. There are many such fields in which a further study of the calculus will be an advantage. The course in the elementary calculus in the senior high school will make it possible for the pupil who has taken it to continue further his study of the calculus without the aid of a teacher.

VII. The seventh topic will give suggestions for a modern presentation of the elementary calculus. These suggestions will show how to simplify the subject, make the topics definite, and lead the pupils to see the theory as a related whole.

Among the sources from which some of these suggestions have been derived is the historical survey of the calculus in the development of mathematics. A comparison of textbooks on the elementary calculus which have been written for beginners has also furnished a means for finding out some of the most promising methods.

An additional source for these suggestions is the author's own experience. For several years he gave a course in advanced algebra to the scientific group in the public high school of a third-class city in which was included a considerable amount of work in differentiation. This work in differentiation included the greater part of the course in the differential calculus for use in the senior high school as it is outlined in the discussion under this section. His experience in teaching the calculus to classes in college has enabled him to determine the explanations necessary to bridge over the gap between the student's understanding and the usual method of presenting the calculus in college. The general training that he received in teaching pupils of a different race and language in the Orient has also been an aid to him in making troublesome topics clear to American pupils.

The author's graduate studies in pure mathematics have furnished him with a criterion for developing the course in the elementary calculus from the standpoint of mathematical rigor; that is, the deductions made throughout the course are sound, although their proofs may not be given. On the practical side, he has had the benefit derived from the study of the mathematics included in a college course in electrical engineering. This benefit has been increased by his association with managers of a large steel plant and of a large electrical firm.

The author is also indebted for suggestions to co-workers at Teachers College to whom he had submitted his manuscript.[2] He also profited by the Round Table Conference in Mathematics where he presented a part of this manuscript.[3]

[2] Among these are Messrs. M. A. Norgaard, assistant professor of mathematics, Grinnell College; J. R. Clark, editor of *The Mathematics Teacher;* W. W. Rankin, professor of mathematics, Agnes Scott College, Georgia; R. Beatley, of the department of mathematics of the Horace Mann School for Boys; and Misses H. Calkins, of the faculty of mathematics of Knox College; P. Wiggins, of the Los Angeles City Schools; T. Schlierholz, of the Ben Blewett Junior High School, St. Louis; R. S. Nichols, of the Vail-Deane School; M. A. Weber, of the Saginaw East Side High School, Michigan.

[3] Round Table Conference, Department of Mathematics, Teachers College, Columbia University, March 19, 1921.

I

A STUDY OF THE STATUS OF MATHEMATICS IN THE SCHOOLS ABROAD THAT CORRESPOND TO OUR HIGH SCHOOLS

AS INDICATED BY THEIR TEXTBOOKS AND SCHOOLS

The status of mathematics in the schools abroad that correspond to our high schools will be indicated principally by a discussion on the extent of work covered by these schools in the subjects of analytic geometry and the calculus. A comparison of the leading European textbooks on the elementary calculus shows that this subject is taught to younger pupils than is customary in our American schools. (A more extended discussion on a comparison of these textbooks is given in Section V.) The German textbooks on the elementary calculus which have been written for use in the Prussian secondary schools adhere rather closely to the scientific method. In France, the elementary calculus forms a part of their course in algebra. The French stress generalization rather than application. However, the English authors of this type of book have made a radical attempt to treat the subject from the point of view of the young student. The author of one of these recent English textbooks, who is himself a teacher of mathematics for engineering students, says in the preface, "Fortunately, the calculus is now beginning to be taught to boys of fifteen or sixteen in day and evening technical schools, and it will grow up with them in the same way as their trigonometry and arithmetic have done." [1]

In the Manchester Grammar School (England) the curriculum for the Mathematical Sixth includes algebra, trigonometry, analytic geometry, and differential and integral calculus. This work corresponds to the work in mathematics given in the freshmen and sophomore years in American colleges.[2]

In the *Lycées* (France), the course in secondary instruction covers a period of seven years and is coördinate with the preceding pri-

[1] John Stoney, *An Introduction to the Differential and Integral Calculus.* (Junior Day Technical School and Municipal Technical School, Smethwick, London.)

[2] *Classical and Science Lists of The Manchester Grammar School, for the Midsummer, 1920.* Furnished by Dr. I. L. Kandel, of Teachers College, Columbia University.

mary instruction which covers a period of four years.[3] The period of secondary instruction is divided into two cycles; the first cycle consists of four years and the second one of three years. In the first year (which corresponds to our eleventh school year) of this second cycle, the course in algebra for the *Latin-Sciences* and the *Sciences-Langues vivantes* sections includes the following work in the calculus: the notion of a derivative, geometric signification of the derivative, the sense of the variation indicated by the sign of the derivative, applications to very simple numerical examples and in particular to functions in preceding studies. In the next year in these two sections, the calculus is extended and includes the calculation of derivatives of the simple functions and a study of rectilinear motion by means of the theory of derivatives. In the last year, the work in the classes in mathematics includes the derivative of a sum, product, quotient, of the square root of a function, of sin x, of cos x, of tan x, and of cotg x, the topics of maxima and minima, and of the area of a curve regarded as a function of the abscissa. The calculus in the special class in mathematics is equivalent to that of our usual college course in the subject.

The reform movement in mathematics in the *Oberstufe* of Germany had the function concept as its central idea.[4] The following divisions into groups shows how the schools reacted toward this idea.

a. Those schools which declined to adopt the idea of the function concept.

b. Those schools which believed in a late discrete development of the function idea: without the infinitesimal calculus; with the differential calculus; with the differential and integral calculus.

c. Those schools which believed in a gradual, continuous development of the function concept: without the infinitesimal calculus; with the differential calculus in the *Prima* (the last year); with the differential and integral calculus already in the *Obersekunda* (the second half of the year preceding the last year).

The nature of the work in the calculus in the *höheren Schulen* in Germany is given under the next topic.

[3] Delalain Frères (ed.), *Plan D'Études et Programmes D'Enseignement dans les Lycées et Collèges de Garçons*, p. xviii, Paris, 1907–1908. F. E. Farrington, *French Secondary Schools*, New York, 1910.

[4] Walter Lietzmann, *Die Organisation des Mathematischen Unterrichts an den Höheren Knabenschulen in Preussen*, p. 199, Leipzig, 1910.

AS INDICATED BY THE COURSES IN ANALYTIC GEOMETRY AND THE CALCULUS IN DIFFERENT COUNTRIES

This discussion is based upon the reports of the International Commission on the Teaching of Mathematics. This Commission owes its origin to the International Congress of Mathematicians at Rome in 1908. Thus far eighteen countries have made reports. These are so extensive that they cover over 12,000 pages.[5]

The work in these two courses, analytic geometry and the calculus, in the foreign schools for the years that correspond to our tenth, eleventh, and twelfth school years will now be considered. The material for this comparison has been selected from the United States Bureau of Education Bulletin No. 45, which gives in systematic form that part of the report of the International Commission on the Teaching of Mathematics that relates to curricula in mathematics.[6]

THE YEAR CORRESPONDING TO OUR TENTH SCHOOL YEAR

The ages of the pupils in this school year (10) are approximately fifteen to sixteen years. In this year in the *Realschule* of Russia, both plane analytics and the elements of the infinitesimal calculus are included in the course in mathematics. In analytic geometry, the principal theorems are derived by means of rectangular coördinates, and the equations for the circle, ellipse, parabola, and hyperbola are derived both in rectangular and polar coördinates. Their infinitesimal calculus is of the nature of an introductory course in the calculus. The simpler formulas for the derivative are taught. The work also includes the topics of maxima and minima, and indefinite and definite integrals.

In this year (10) the analytic geometry is also taught in the *Realgymnasium* and *Oberrealschule* of Germany.

THE YEAR CORRESPONDING TO OUR ELEVENTH SCHOOL YEAR

The ages of the pupils in this school year (11) are approximately sixteen to seventeen years. An introductory course in analytic

[5] Raymond Clare Archibald, *The Training of Teachers of Mathematics*, United States Bureau of Education Bulletin, No. 27, 1917.

[6] J. C. Brown, *Curricula in Mathematics*, United States Bureau of Education Bulletin, No. 45, 1914.

geometry is given in this year (11) in the *Gymnasium* in Austria, and a more difficult course in the *Realschule*. In the latter school, easy differential and integral calculus are introduced in order to simplify or make more intensive the knowledge of physics.

In this year (11) in the *Oberrealschule* in Germany, courses in analytic geometry and in the differential calculus are included in the work in mathematics, but the calculus is not taught in this year in the *Gymnasium* and the *Realgymnasium*. A course in analytic geometry is also included in this year in the curricula of the three main schools of Austria, of the *Lycées* of France, and of the *Realschule* and *Gymnasium* of Hungary, and of the *Realgymnasium* of Sweden.

THE YEAR CORRESPONDING TO OUR TWELFTH SCHOOL YEAR

The ages of the pupils in this school year (12) are approximately seventeen to eighteen years. In this year, analytic geometry is taught in the *Gymnasium* schools of Holland, and of Hungary. Also, in this school year which corresponds to our twelfth school year, both analytic geometry and the calculus are taught in Austria, Belgium, Denmark, England, France, Germany, Roumania, Sweden, and Switzerland. The work in these two subjects ranges from that of a simple introduction to a course equivalent to our first course in college.

CONCLUSIONS

Thus, it is evident that the United States is far behind other countries in the work that our high schools do in mathematics.[7] That this difference has not appeared more prominently heretofore in such fields as engineering is partly due to the fact that our larger quantity of superior native ability has been able to compensate this difference. Owing to the rapid expansion of all fields making use of mathematics, however, and to the stronger competition resulting from the closer relations of the nations of the world both as to the shortening of time in transportation and as to political interests, the need of more men with trained mathematical minds cannot be met by relying upon native ability alone but by giving this native

[7] Even the conditions in our own colleges show that the calculus should be placed earlier in the curriculum since frequently engineering students are asked to apply its principles in physics and mechanics before these principles have been developed in the regular calculus course.

ability mathematical training of the most careful sort. To give this training in a satisfactory manner, it must be begun in the senior high-school period.[8]

The main reason that pupils in the schools abroad are able to accomplish this work in mathematics which throws them about two years ahead of our American pupils by the end of the twelfth school year is because of the more profitable arrangement of the courses in the mathematics curriculum in their schools. A more profitable arrangement has already been made in the mathematics curriculum of the American standard junior high school, and an attempt is now being made to do a similar thing for the mathematics curriculum of our senior high school. In the junior high school, the propædeutic courses in algebra, geometry, trigonometry, and graphs lay the foundations for the senior high-school courses in additional algebra, demonstrative geometry, additional trigonometry, and the introductory work in analytic geometry. Thus, the foundation is laid for an introductory course in the elementary calculus in the twelfth school year. A scheme such as this would furnish an opportunity to our pupils who have the inclination and the native ability to compete favorably with the highly trained pupils abroad, and would aid us as a nation in maintaining a foremost place in the fields of applied science and of engineering.

[8] "If the student who omits the mathematical courses has need of them later, it is almost invariably more difficult and it is frequently impossible for him to obtain the training in which he is deficient. In the case of a considerable number of alternative subjects a proper amount of reading in spare hours at a more mature age will ordinarily furnish him the approximate equivalent to what he would have obtained in the way of information in a high-school course in the same subject. It is not, however, possible to make up deficiencies in mathematical training in so simple a fashion. It requires systematic work under a competent teacher to master properly the technique of the subject and any break in the continuity of the work is a handicap for which increased maturity rarely compensates. Moreover, when the individual discovers his need for further mathematical training, it is usually difficult for him to find the time from his other activities for systematic work in elementary mathematics."—Elective Courses in Mathematics for Secondary Schools, a preliminary report by The National Committee on Mathematical Requirements.

II

THE TREND OF MATHEMATICS IN OUR PUBLIC SCHOOL SYSTEM

A REASON FOR THE REORGANIZATION IN THE ELEMENTARY-HIGH-SCHOOL PERIOD

The mathematics curriculum of the first nine school years has already adapted itself to the changed conditions due to the reorganization in the elementary-high-school period. "One of the main reasons for taking two years from the elementary-school period and adding them to the high-school period, thus dividing the elementary-high-school period of twelve years into a period of six years of elementary school and six years of high school, and then subdividing the high-school period into three years of junior high school and three years of senior high school, was to make possible a better adjustment of studies and school methods in the three years now coming to be quite commonly included in the junior high school." [1]

JUNIOR HIGH-SCHOOL MATHEMATICS

The nature of the mathematics in the three years of the junior high school can probably be most simply illustrated by an outline of the contents of a suitable set of textbooks on junior high-school mathematics. There are a number of such sets of textbooks on the market. The following is the outline of the contents of one of these:

FIRST YEAR

- First half. Arithmetic.
  - Arithmetic of the home, the store, the farm; arithmetic of industry; arithmetic of the bank; material for daily drill.
- Second half. Geometry.
  - Geometry of form; geometry of size; geometry of position; supplementary work; tables for reference.

SECOND YEAR

- First half. Algebra.
  - The formula; the equation; the graph; negative numbers; algebraic operations; further uses of algebra.

[1] United States Bureau of Education, Secondary School Circular, No. 6, July, 1920. Statement by Commissioner P. P. Claxton.

Second half. Arithmetic.

Arithmetic of trade, of transportation, of industry, of building; arithmetic of the bank, of corporations; arithmetic of home life, of the farm, of community life, of civic life; arithmetic of investments, of mensuration; material for daily drill; tables for reference.

THIRD YEAR

First part. Algebra.

The chief uses of algebra; addition and subtraction; multiplication; division; fractions; simple equations; quadratic equations; general review.

Second part. Trigonometry.

Functions of angles, with applications; trigonometric tables, with applications.

Third part. Demonstrative geometry.

Meaning of geometry; triangles; parallel lines; parallelograms; angle sums; areas; constructions; loci; review; table for reference.[2]

According to the arrangement of the work in mathematics in the standard junior high school, propædeutic courses have been given in algebra, geometry, and trigonometry by the end of the ninth school year. The nature of the subject-matter in these three branches differs from the customary textbook type. Instead of an unnecessary amount of time and space being given to puzzle problems, problems of mere complexity, and problems of a too advanced character merely to make the work exhaustive, the time and space are devoted to a simple, careful, and helpful introduction into these subjects.

Now that the mathematics curriculum of the junior high school has been made definite, it can be used as a basis upon which to build the mathematics curriculum of the senior high school.

SENIOR HIGH-SCHOOL MATHEMATICS [3]

Mathematics Elective. It cannot be expected nor would it be advisable that mathematics should be required of all pupils in the senior high school. One of the reasons for this is that there are some pupils who have not minds capable of the amount of rigorous thinking required in mathematics. These pupils will not elect mathematics, and those who do elect mathematics will do so largely because they like the subject. For this reason, there will not be the need of arti-

[2] Wentworth-Smith, *Junior High-School Mathematics*, New York.

[3] Some of these ideas were expressed by Professor D. E. Smith in his lectures at Teachers College, February, 1921.

ficial motivation. Hence, we can follow the shorter road into mathematical insight by making use of suitable material that may not have any practical application whatever, and the pupils will enjoy it. There will not then be the necessity for adapting mathematics to mathematically weak minds since these pupils will not take mathematics. This selection will give us the stronger type of mathematical mind and that strong kind of incentive which comes from pupils who have mental hunger for this particular line of study. These pupils will take up the subject for the love of it. Accordingly, so far as mathematics is concerned, our high schools will be better off than corresponding schools abroad, since there the selection of pupils is usually on the basis of wealth and not on the basis of ability.

Various Courses. Various courses in which this elective mathematics may run are vocational, commercial, industrial, and agricultural courses. Commercial mathematics is such an extensive subject that it is taken care of by itself in the commercial high schools. There this work covers four years. There is also a great deal of mathematics required in industrial work, as is shown by the numerous technical schools which exist. There is little mathematics needed in agriculture. Also, there will be some pupils who will want mathematics for informational purposes and some for college entrance.

The possible courses themselves will now be taken up according to the school years in the senior high school.

The Tenth School Year. In the tenth school year, what should the student have if he likes mathematics, or if he wants to go to college, or if he wants to enter industrial fields where he will find it advantageous to continue further his studies in mathematics of his own account? There is no question but that he will need to take additional work in algebra. This too would be the time for him to become further acquainted with those geometric ideas which later will make it possible for him to grasp more readily the mathematical conditions involving such ideas. At the same time, the course in plane geometry should be so arranged that it will serve one of its main purposes, namely, to furnish the best developed applications of logic that we have. Along with this work in plane geometry, there should be given at opportune times the related notions of solid geometry in order that the learner's mind may not be unnecessarily restricted to a geometric world of two dimensions. Such an arrangement will also give the preliminary exercise in the development of three dimensional space perception. In fact, there is no longer a necessity for teaching solid geometry as a separate subject in our high schools, and there is a tendency in a number of our high schools toward the elimination of solid geometry as a separate subject.[4] Its elimination can be effected without the loss of its three main aims, the development of space perception, the extensive work in mensuration, and its introductory and almost intuitive treatment of incommensurables and limits. It has already been suggested that the first of these can be partly taken care of in the course in plane geometry. The more extended practice in the development of space perception, the extensive work in mensuration, and the treatment of incommensurables and limits can be presented in a much more satisfactory manner in a course in the elementary calculus adapted for use in the senior high school.

Hence, the two main courses in mathematics in the tenth school year should be algebra and demonstrative geometry. With the propædeutic courses in mathematics in the junior high school as a foundation, the minimum time to be devoted to either of these two subjects need not be greater than half a school year.

The Eleventh School Year. The first half of this year could be devoted to elementary courses in statistics and descriptive geometry. This course in elementary statistics would be a valuable aid to the pupil in his general reading. It would enable him to comprehend more readily the concentrated information contained in statistical charts that are found so frequently in current articles. The course in descriptive geometry would aid the pupil in extending his knowledge concerning the properties of solids and develop skill in drawing in perspective. This kind of drawing is the one most used in future work.

The second half of the year could be devoted to work in trigonometry in addition to the numerical trigonometry given in the junior high-school mathematics. In the second half of this year, there should also be given a considerable amount of graph work. The finding of maxima and minima of quantities by means of graphs should be especially emphasized. In rectangular coördinates, the

[4] *Questionnaire to Determine the Place of Solid Geometry in the High-School Curriculum*, sent out by the author, spring, 1920.

equations and some of the simple properties of the straight line, the circle, the parabola, the ellipse, and the hyperbola should be taught.

The Twelfth School Year. It is seen that there is room in the last year of the senior high school for an additional course in mathematics. The mathematics of the tenth and eleventh school years were preparing the pupil principally to enter one of two fields, either the field of pure mathematics or some special field of industry in which he would find it desirable to continue further some of his studies in mathematics. Is there one such course that will give the pupil desiring to enter the industrial field such additional introduction into the field of mathematics that he will be able thereafter to continue his further study in mathematics without the aid of a teacher? Some of the possibilities of such a course will now be discussed.

In the twelfth school year, what may be done for the pupil from the standpoint of his love for mathematics and of the use it will be to him in his general reading? He might be given a more extended course in analytic geometry. This latter course does not lead anywhere in this school year except to analytic geometry since there are no applications. He might be given higher algebra, including such topics as permutations and combinations, theory of equations, and the theory of determinants. It is less than one hundred years that people have been able to multiply one determinant by another. However, higher algebra also does not lead anywhere. There might be included a course in projective geometry. This would lead the pupil up upon a high mountain so far as the field of geometry is concerned, and this would be very interesting. From the geometric side alone this is one of the best things that could be done but it too has no applications, and hence for a good many minds it would not be as interesting as something that has some applications. Yet there is a subject that is interesting, that is not too difficult, that leads to work having applications, that has an important propædeutic value, and that will deepen the pupil's appreciation of mathematics. This subject is an introductory course in the elementary calculus. Accordingly, in the last year of the senior high school, or the twelfth school year, there should be offered as an elective in mathematics an introductory course in the elementary calculus. This raises the question, are there a sufficient number of teachers available who are capable of teaching this course?

A Sufficient Number of Teachers Available. The question of obtaining a sufficient number of teachers qualified to teach the elementary calculus in the senior high school does not present much difficulty. There are a number of teachers now in the service who can do this. The later standards of qualifications for teaching will increase this number. An illustration of one of these standards is as follows: "The minimum attainment of teachers of any academic subject shall be equivalent to graduation from a college belonging to the North Central Association of Colleges and Secondary Schools requiring the completion of a four year course of study or 120 semester hours in advance of a standard four year high-school course." [5]

Such standards as the one just quoted will cause the prospective teacher of mathematics to include sufficient mathematics in his preparatory work to qualify him to teach the elementary calculus of the senior high school. Also, there is the opportunity of teachers now in service to prepare themselves for this work by taking work in mathematics in summer sessions. "Summer schools are conducted by universities and colleges and normal schools, or are specially organized by State educational authorities. In general, their work is intended for teachers in service. Courses in mathematics have been regarded in most cases as of special importance in these summer curricula. In fact, in several instances the work began with courses in mathematics and perhaps one other subject, and was extended to include courses in all the regular departments as the number of students increased and the demand became apparent." [6]

The following university standard is additional evidence that soon there will be no lack of teachers of mathematics qualified to teach the elementary calculus in the senior high school: "The department of mathematics at Brown University lays down a minimum course of study which it requires students who are prospective

[5] *Report of North Central Association of Colleges and Secondary Schools*, edited by C. O. Davis, p. 37-A, 1920. This association also reports seventy-seven bona fide senior high schools. This report was obtained through the aid of Provost C. B. Upton, of Teachers College.

[6] *The Training of Elementary-School Teachers in Mathematics in the Countries represented in the International Commission on the Teaching of Mathematics*, I. L. Kandel, United States Bureau of Education Bulletin, No. 39, p. 53, 1915.

teachers of mathematics to take if they wish the backing of the department in starting on their careers. In outline the course is as follows: Plane trigonometry (3 semester hours), higher algebra (3), solid geometry (3), plane analytic geometry (4), differential and integral calculus (8), teachers' course in algebra (6), and teachers' course in geometry (6)." [7]

Even the prospective junior high-school teacher of mathematics should realize that he will need considerable preparation in pure mathematics if he wishes to obtain a proper appreciative attitude toward the elementary content of the junior high-school curriculum in mathematics.[8] The prospective junior high-school teacher needs to have strengthened or even first imparted the following concepts, skills and information: formula work, graph work, concept of the function, trigonometry, computation, drawing, and properties of solids. For his basic foundation for formula work, he should go to trigonometry; for graph work, to analytic geometry; for the concept of the function, to the calculus; for computation, to logarithms; for drawing and the properties of solids, to descriptive geometry.

That the number of teachers who are college graduates is on the increase is indicated by the following quotation from statistics for the State of New Jersey for the year ending in June, 1920. "In high schools the number of new teachers who were graduates of colleges, universities or technical institutions was 89 more than last year." [9]

CONCLUSIONS

The discussion in this topic has shown that the courses in mathematics preceding the twelfth school year not only form a suitable preparation for but also leave room for the introduction of the elementary calculus in the twelfth school year.

This discussion also shows that there will be available a sufficient number of teachers qualified to teach the elementary calculus in the senior high school, since there are a number of such qualified

[7] Bureau of Education Bulletin, No. 27, by Raymond Clare Archibald, Associate Professor of Mathematics in Brown University, p. 206, 1917.

[8] Percival M. Symonds, "Subject Matter Courses in Mathematics for the Professional Preparation of Junior High-School Teachers," *Educational Administration and Supervision*, February, 1921.

[9] Furnished by Professor T. H. Briggs, Teachers College, Columbia University.

teachers now in the service, since others will take the opportunity to prepare themselves for teaching the elementary calculus, and since the future teachers of mathematics will have sufficient mathematics included in their preparatory work to qualify them for this purpose.

III

THE IMPORTANT POSITION OCCUPIED BY THE CALCULUS IN THE MATHEMATICS STRUCTURE

Practically all the mathematics that everybody should know is included in the curriculum of the first nine school years. There is also some elective mathematics offered in the ninth year, or last year of the junior high school. It was shown in the discussion under the preceding topic that upon this foundation there should be built the courses in demonstrative geometry, algebra, and trigonometry; and, also, that these courses together with the work in graphs form a suitable preparation for the simpler parts of the elementary calculus in the twelfth school year.

It will now be well to investigate what lies beyond the twelfth school year in the field of mathematics. There are two main divisions in this later field, namely, pure mathematics and applied mathematics. Pure mathematics includes such subjects as higher calculus, differential equations, functions of a complex variable, integral equations, group theory, modern higher algebra, theory of numbers, and Einstein's theory of relativity. The first four and the last of these subjects make continual use of the calculus.

Applied mathematics presents a yet wider range of subjects which make use of the calculus. Some of these are the mathematics of mechanics, of physics, of electricity, of light, of ship-building, of statistics, of aviation, of wireless telegraphy, and of engineering. Those who are familiar with the courses in engineering know that the calculus contributes so much to its theory that much of our engineering work would be impossible without the calculus. A similar thing may be said of mechanics, physics, and electricity. In the last named subject, the calculus is indispensable for a thorough exposition of the phenomena of alternating currents.[1] A knowledge of the calculus is also necessary for an exhaustive study of the remaining subjects. Even an elementary textbook on wire-

[1] Charles Proteus Steinmetz, *Theory and Calculation of Alternating Current Phenomena.* New York, 1916.

less telegraphy contains some calculus.[2] To-day, the aeroplane is built according to the knowledge obtained by the careful methods of the laboratory, and the current numbers of aviation journals frequently contain articles which cannot be understood without a knowledge of the calculus.[3]

The wide application of one of the topics in the calculus is shown by the following quotation. "As examples of the utility of these theorems (maxima and minima) may be cited the finding of the positions and magnitudes of maximum bending moments, of maximum stresses, of maximum deflections, of maximum velocities, of maximum accelerations of momentum, of the positions of rolling load on bridges to give maximum stress in any given member of the bridge, etc., etc. All these things are of special importance in the practical theory of engineering. In the jointing of pieces in machines and static structures, it is never possible to obtain uniform stress over the various important sections of the joint. It is of the greatest importance to find the maximum intensities of stress on such sections, because the safety of the construction depends upon the maximum, hardly ever upon the average, stress. The average stress on the section is found by dividing the whole load on the section by the whole area of the section. Such average stresses are often very different from the maximum stress, and no reliance ought to be placed upon them as measures of strength and safety." [4]

CONCLUSIONS

Thus it is seen that the calculus is a most important aid in applied mathematics and contributes largely to the field of pure mathematics. It is also evident that the elementary calculus is *the* connecting link between the fields of pure and applied mathematics *and* the school work in the proposed curriculum in mathematics preceding the twelfth school year.

Hence, because of its importance, twelfth year pupils in mathematics should be given an opportunity to become acquainted with the elementary calculus, provided it is not too difficult.

[2] S. J. Willis, *A Short Course in Elementary Mathematics and Their Application to Wireless Telegraphy*, London, 1917.

[3] *Flight*, Sept. 2, 1920, Official organ of the royal aero group of the United Kingdom; *Aeronautics*, Sept. 2, 1920, London; *The Aeronautic Journal*, Sept., 1920; *Aircraft Engineering*, Jan., 1920, London.

[4] Robert H. Smith, *The Calculus for Engineers and Physicists*, p. 106, London, 1897.

IV

A HISTORICAL SURVEY OF THE NATURAL GROWTH OF THE CALCULUS IN THE DEVELOPMENT OF MATHEMATICS

REASONS FOR MAKING THIS HISTORICAL SURVEY

Different authors of textbooks on the calculus use different methods in their presentation of the subject. Not all of these methods would prove satisfactory in teaching the elementary calculus to senior high-school pupils. Which of these methods may prove satisfactory? Or what parts of these methods? Or are there additional methods that would prove satisfactory? An aid in answering these questions is the light of past experience. Accordingly, a reason for giving a historical survey of the natural growth of the calculus in the development of mathematics is that it may serve as a guide in the evaluating of methods of presentation of the calculus.

In many ways the mental development of the child follows the mental development of the race. Then a study of the way in which the calculus developed in the minds of men throughout the ages will suggest methods of how it should now be taught to younger pupils. Such a study will point out many of the difficulties that arose in the development of the calculus and how these difficulties were overcome. It will show some methods that are simple and satisfactory and others that are cumbersome and undesirable.

This historical survey will add to the appreciation of the subject because it will show that we could not be enjoying the fruits of the calculus as the result of its application to the fields of science and engineering had it not been for the labors of many mathematicians throughout many centuries. This survey will also show that an introductory course in the elementary calculus is not too difficult for younger pupils. Since the introductory ideas of the calculus are within the comprehension of younger pupils, the appreciation of the subject is an added reason for introducing the elementary calculus into the senior high-school mathematics.

The chronological order has been followed in this historical survey. A representative mathematician has been selected in a convenient span of years, and some of his contributions have been discussed in order to show the status of the development of the calculus at that particular time.

THE PERIOD PRECEDING ISAAC BARROW

The First General Step in the Development of the Calculus. The first step in the development of the calculus is found among the Greeks, and is known as the method of exhaustion. This method was used by Antiphon (*c.* 430 B.C.) in showing that the area of a circle is the limit of the area of an inscribed regular polygon as the number of sides of the polygon is increased indefinitely.[1]

Antiphon's first method of procedure was that of inscribing a square in a circle. Then he doubled the sides of the square. Then he doubled the sides of the resulting polygon, and so on. However, Bryson, who lived about this time, was not satisfied with merely using an inscribed polygon; he also used a circumscribed polygon. His conclusion was that the area of the circle is the limit of the area of the variable inscribed polygon and also that of the variable circumscribed polygon.

These illustrations furnish two ideas in the foundation of the calculus. One of these is the idea of a limit. The other is the idea of the division of a quantity carried on indefinitely. Both these ideas are presented to senior high-school pupils in their course in demonstrative geometry. Hence, if any of these pupils elect the course in the elementary calculus in their last year in the senior high school, they will already have become acquainted with the first step in the foundation of the calculus.

A Bit of Integration. There was a near approach to integration considered as a process of summation by Archimedes in his method for finding the area of a parabolic segment. He made use of the summation of a series: A, $\frac{1}{4}A$, $(\frac{1}{4})^2 A$, . . . , and recognized that $(\frac{1}{4})^n$ approaches zero.[2]

Here, again, the pupils who have taken the work in the mathematics preceding the course in the elementary calculus have already

[1] David Eugene Smith, ***History of Mathematics***, Chap. on The Calculus, Boston, 1921.
[2] *Ibid.*

become acquainted with this method in the proof of the following proposition in geometry: The area of a parabolic segment made by a chord is two-thirds the area of the triangle formed by the chord and the tangents drawn through the ends of the chord.

The Second General Step in the Development of the Calculus. The second general step was taken two thousand years later, and is the method of infinitesimals. This method was used by both Kepler (1616) and Cavalieri (1635).

Kepler, too, had performed a crude kind of integration (1609). He found that "a planet describes equal focal sectors of ellipses in equal times." Expressed in our present notation this would be $\int_0^a \sin k\, dk = 1 - \cos k$, where k is the angle of rotation. In the wine-year of 1612, Kepler was confronted with the problem of giving practical rules for finding the contents of wine-casks. He thought out a method by infinitesimals which he applied in measuring the volumes of solids of revolution. He divided areas and volumes into elementary parts and then found their sum. Without a fully developed theory of infinitesimals, he did this by intuition where the theory lacked. His idea of division into elementary parts enabled him to find the form of a cask which would give the maximum volume for the material used in its construction. He also knew how to test an expression for a maximum value. This test was that the difference between the maximum itself and the immediately preceding and the immediately following values will be imperceptible if the range covered is made small enough.[3]

This historical sketch thus far shows that the integral calculus was developing by itself and not as the inverse of differentiation. Later, in discussing methods of presentation of the calculus, the problem of the integral calculus will be introduced by itself and then it will be shown that integration is the inverse of differentiation.

This account of the contribution of Kepler to the calculus shows that the idea of infinitesimals may be grasped and used without previous training in a rigorous treatment of the theory of infinitesimals. Also, his test for a maximum is both simple and satisfactory, and a method similar to it is included in the elementary course of the calculus for use in the senior high school.

[3] Heinrich Wieleitner, *Geschichte der Mathematik*, II Teil, 1 Hälfte, p. 107, Leipzig, 1911.

BARROW, NEWTON, AND LEIBNIZ

Isaac Barrow. Barrow considered the motion-concept as the main idea in his investigations. In one case he derived the path of a moving point referred to time and velocity, and in another case he derived the velocity of the motion referred to time and path. By considering these two ways, Barrow was the first to consider integration and differentiation as inverse problems.

Barrow's differential triangle method will be given here, since the method used in the elementary calculus for the senior high school for explaining the derivative by means of a geometrical representation is based upon it. His method is as follows:

"Let *AP*, *PM* be right Lines given in Position (whereof *PM* cuts the proposed Curve in *M*,) and let *MT* touch the Curve in *M*, and cut the right Line *AP* in the Point *T*. Now to determine the length of the right Line *PT*, I suppose the Arch *MN* of the Curve to be indefinitely small, and draw the right Lines *NQ*, *NR* parallel to *MP*, *AP;* I call *MP*, *m; PT*, *t; MR*, *a; NR*, *e;* and give Names to other Lines useful to our purpose, determined from the particular Nature of the Curve; and then compare *MR*, *NR* expressed by Calculation in an Equation, and by their means *MP*, *PT* themselves; observing the following Rules at the same time.

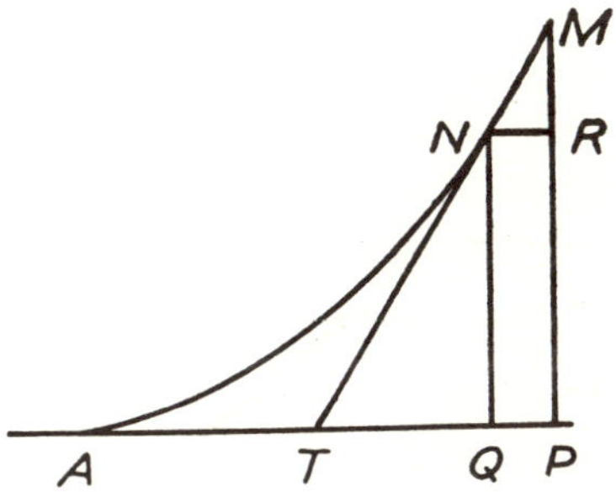

Fig. 1

1. I reject all the Terms in the Calculation, affected with any Power of *a* or *e*, or with the product of them; for these Terms will be equal to nothing.

2. After the Equation is formed, I reject all the Terms wherein are Letters expressing constant or known Quantities; or which are not affected with *a*, or *e;* for these Terms brought over to one side of the Equation will be always equivalent to nothing.

3. I substitute *a* for *m* (*MP*), and *t* (*PT*) for *e;* by which means the Quantity of *PT* will be found.

When any indefinitely small Particle of the Curve enters the Calculation, I substitute in its stead a Particle of the Curve properly taken; or any right Line equal to it, because of the indefinite Smallness of the Part of the Curve."[4]

By means of his differential triangle method, Barrow could find the tangent to a curve at any point. He lacked a classified set of formulas which would have facilitated this work. Also, there was at this time no rigorous treatment of the theory of limits. This method of Barrow's justifies the presentation of the derivative as the limit of a geometric ratio to pupils who possess only an intuitive and elementary knowledge of limits.

Isaac Newton. The third general step in the development of the calculus was the invention of fluxions by Newton (*c.* 1665).[5]

The method of fluxions played a very important part in the development of the calculus. A little insight into the nature of the subject will now be given through a brief explanation of some of the terms and notation as used by Newton.

By a flowing quantity he meant the trace that is being made by a moving point. The velocity of the moving point at any particular place he called the fluxion. He represented the flowing quantities by such letters as x, y, z, and their fluxions by placing a dot above the respective letters, as $\dot{x}$, $\dot{y}$, $\dot{z}$. He represented infinitely small increases or decreases by putting a small zero before the letter. His rules for differentiation were similar to the ones used by Barrow.[6]

In our present-day textbooks on the calculus, the topic of rates resembles the method of fluxions. It is seen that the method of fluxions requires a considerable knowledge of physics. For this reason, the elementary calculus based upon infinitesimals and geometrical representations is more suitable for our American pupils.

Although the notation of Newton was zealously adhered to by a number of the English mathematicians for over a hundred years, it finally was replaced by the more convenient notation of Leibniz. The Analytical Society in 1820 published two volumes using Leibniz' notation for the calculus; one was by Peacock and the other by Herschel. These volumes were books of examples illustrating the new notation. Since then this notation has been used by all elementary textbooks.[7]

[4] *Geometrical Lectures by Isaac Barrow*, Translated by Edmund Stone, pp. 173, 174, London, 1735. These Geometrical Lectures had been published by Barrow in 1670.

[5] David Eugene Smith, *History of Mathematics*, Chap. on the Calculus, Boston, 1921.

[6] John Wallis, *De Algebra Tractus*, pp. 391, 392, 393, Oxonia, Ed. 1693.

[7] W. W. R. Ball, *History of Mathematics at Cambridge*, p. 122, 1889.

Gottfried Wilhelm Leibniz. The following quotation states tersely a fundamental difference between the calculus of Leibniz and that of Newton. "Leibniz did not reply to this letter [Newton's about inverse tangents], until June 21, 1677. In his answer he explains his method of drawing the tangents to curves which he says proceeded 'not by fluxions of lines but by differences of numbers'; and he introduced his notation of dx and dy for the infinitesimal differences between the coördinates of two consecutive points on a curve. He also gives a solution to the problem to find a curve whose subtangent is constant which shows that he could integrate." [8]

Leibniz' idea of an infinitesimal as a minute difference between the coördinates of two consecutive points is simpler than Newton's fluxion idea because it does not involve the notion of velocity and a knowledge of the laws of physics. Because of its greater simplicity, the idea of infinitesimal differences is used as the basis of the senior high-school calculus. The notation introduced by Leibniz is also used in the senior high-school calculus in a somewhat simplified form.

THE EARLY TEXTBOOKS ON THE CALCULUS

Jean Bernoulli. Although this work of Bernoulli did not appear in book form, his manuscripts written in 1691–92 form the first integral calculus.[9]

The first lecture concerns the nature and the calculation of the integral. His method is that of the inverse of differentiation. He says that it is known that dx is the differential of x, and xdx the differential of $\frac{1}{2}x^2$ or $\frac{1}{2}x^2 \pm$ a constant quantity, and so on. He then generalizes to some extent and forms the following table:

adx is the differential of $ax \pm$ a constant
$axdx$ is the differential of $\frac{1}{2}ax^2 \pm$ a constant
ax^2dx is the differential of $\frac{1}{3}ax^3 \pm$ a constant
ax^3dx is the differential of $\frac{1}{4}ax^4 \pm$ a constant, and so on.

[8] *Ibid.*, p. 58.

[9] *Die Erste Integralrechnung*, Eine Auswahl aus Johann Bernoullis Mathematischen Vorlesungen über die Methode der Integrale und anderes aufgeschrieben zum Gebrauch des Herrn Marquis de l'Hospital in den Jahren 1691 und 1692 als der Verfasser sich in Paris aufhielt, by Dr. Gerhard Kowalewski, London, 1914. This is the first translation of these manuscripts into German.

From this he deduces the general formula that $ax^p dx$ is the differential of $\frac{a}{p+1} \cdot x^{p+1}$.

Since the pupils in the senior high school are accustomed to inductive reasoning, this first attempt at a table of integrals is a good method for them to follow in their calculus work.

Bernoulli says that direct, important mathematical problems and theorems depend upon the finding of the integral, such as the quadrature of surfaces, the rectification of curves, the cubes of solids, the inverse tangent method or the finding of the nature of the curve from given properties of the tangent, and also such as pertain to mechanics, as the method of finding the center of gravity, of impulses, of vibratory motion, and so on. Also, through the finding of the integral we obtain the evolute of curves, and the art of finding their nature, and with the help of the evolute of rectifying the curves themselves.

This shows the wide use that was made of the calculus over two hundred years ago. Since then its fields of application have been greatly extended in pure mathematics, in science, and in engineering. This usefulness of the calculus is evidence of its importance in the mathematical structure.

Bernoulli's second lecture pertains to the quadrature of surfaces. He says that of the many different parts into which we divide the integral calculus, by far the first and most important is that which presents the quadrature of surfaces. In accordance with this plan, the integral calculus in the elementary course in the senior high school is introduced by giving an illustrative problem in finding the area of a surface under a parabola.

Bernoulli considers the surface as divided into an infinite number of parts, each of which may be regarded as the differential of the surface. If then we have the integral of this differential, that is the sum of these parts, then will we also know the desired area. Those infinitely little parts of the plane surface can be thought of in different ways, according to the most convenient way permitted by the details of the plane. Either the plane surfaces would be directly divided, as is the custom, by means of parallel lines, as in Fig. 2, or by means of an infinite number of straight lines which are concurrent, as in Fig. 3, or by means of an infinite number of tangents, as in Fig. 4, or by means of an infinite number of normals to the curve, as in Fig. 5.

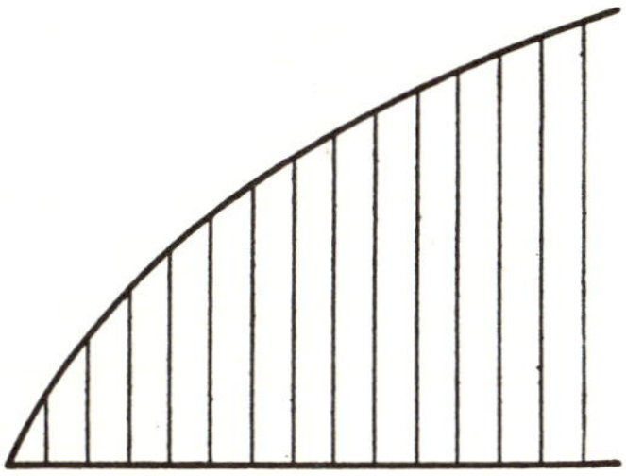

FIG. 2

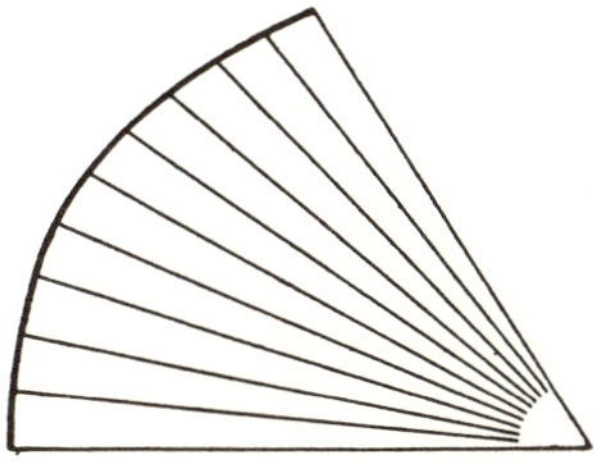

FIG. 3

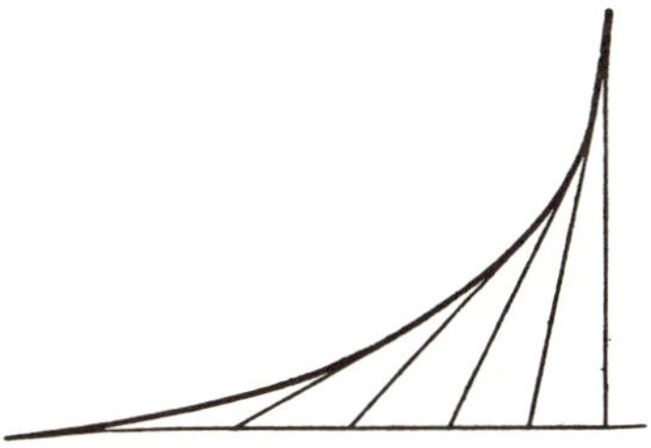

FIG. 4

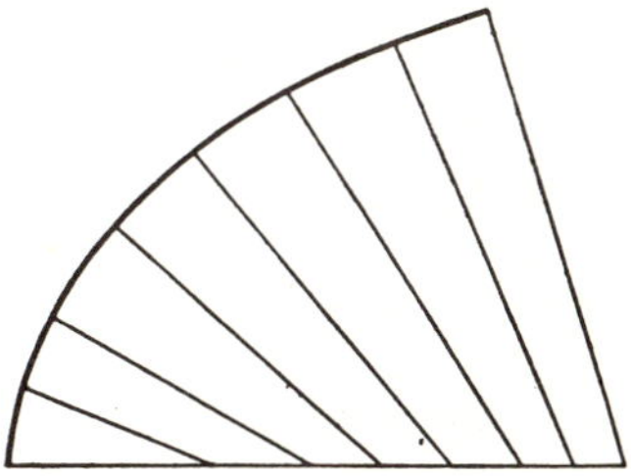

FIG. 5

If the divisions of the surface are parallel, and if x is the abscissa and y is the ordinate, the differential of the surface will be ydx; namely, the rectangle formed by the ordinate and the differential of the abscissa. If AC is the given curve, then y will have a given ratio to x, so that it is expressed in x alone. Suppose that AC is a parabola, then $ax = y^2$ or $y = \sqrt{(ax)}$. The integral of this, which is $\frac{2}{3}x\sqrt{(ax)}$ or $\frac{2}{3}xy$, is the area desired.

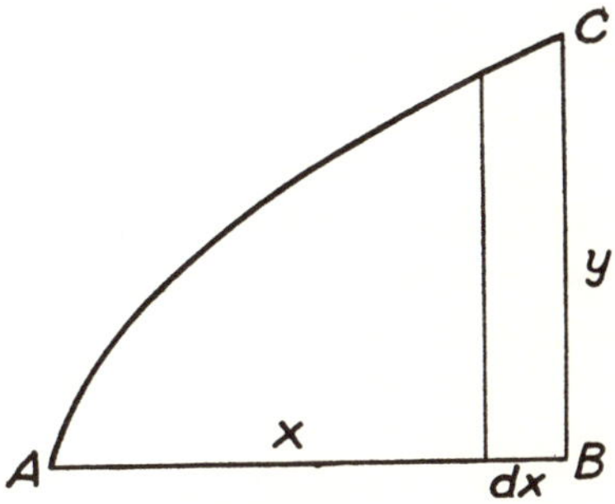

FIG. 6

This method of Bernoulli's for finding the area is simple. It is also satisfactory, although the idea of limits involved is taken care of largely by intuition. This is a justification for the use of this method somewhat modified in introducing the subject of integration as a process of summation in the senior high-school calculus.

Marquis de l'Hospital. This mathematician wrote the first textbook on the differential calculus. The book was published in 1696.[10]

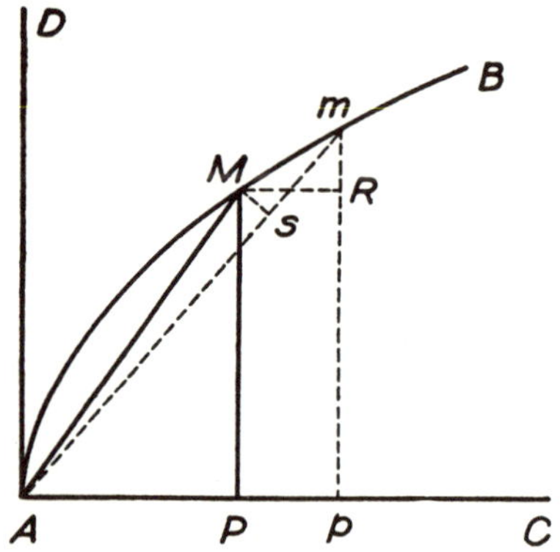

FIG. 7

L'Hospital considers a "difference" as an infinitely small portion of a variable quantity that increases or decreases continually. He gives a number of illustrations of differences. Pp will be the difference of AP; Rm that of PM, Sm that of AM, Mm that of the arc AM, the triangle MAm that of the segment AM, the little space $MPpm$ that of the mixtilinear triangle AP, PM, and the arc AM. If the variable AP is represented by x, then dx expresses the value of Pp, and so on.

This simple method, somewhat modified, is used in the explanation of the increment of a variable in the senior high-school calculus.

Charles Hayes. Charles Hayes published the first textbook on the calculus in English in 1704.[11]

A comparison of different methods used in differentiation at this time can be made to some extent by considering the three methods that Hayes presents for the differentiation of a product.

He states the proposition as follows: "If X be multiplied into Z, and if the Product be XZ, I desire to know the Fluxion of the Rectangle XZ; That is, supposing the Sides X and Z to be augmented or diminished each by an infinitely little Quantity, I would know how much the new Rectangle exceeds or is exceeded by the given Rectangle XZ."

Although Hayes used the fluxional notation, the Leibniz notation

[10] Marquis de l'Hospital, *Analyse des infiniment Petits pour l'Intelligence des Lignes Courbes*, Paris, 1696.

[11] Charles Hayes, *A Treatise of Fluxions, or an Introduction to Mathematical Philosophy*, Gent., London, 1704.

will be used here in the methods he gives since it is more familiar to the present-day reader.

First method. A geometrical figure is used. If X be multiplied by Z, the rectangle is XZ, and we suppose that half the little increment of X to be $\frac{1}{2}dx$, and half the little increment of Z to be $\frac{1}{2}dz$; it is evident that the differential of the rectangle is

$$X \cdot \tfrac{1}{2}dz + Z \cdot \tfrac{1}{2}dx + r.$$

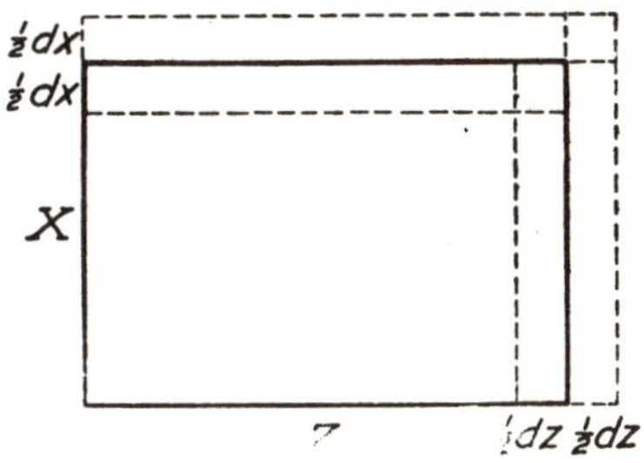

FIG. 8

Again, suppose that half the infinitely little decrement of X and of Z to be $\frac{1}{2}dx$ and $\frac{1}{2}dz$ respectively, then the decrement of the rectangle XZ is equal to $X \cdot \frac{1}{2}dz + Z \cdot \frac{1}{2}dx - r$, and adding the increment and the decrement into one sum, we have $Xdz + Zdx$ for the differential of the rectangle XZ.

The second method. This method makes use of the algebraic process. The "Incomparable Mr. Newton" shows how to find the differential of any rectangle in a manner nothing differing from this, only expressed another way. For instance, to find the differential of the rectangle XZ, he supposes the differential of X and of Z to be dx and dz. Then,

$$\begin{array}{l} X+\tfrac{1}{2}dx \\ Z+\tfrac{1}{2}dz \\ \hline XZ+\tfrac{1}{2}dx \cdot Z \\ \quad +\tfrac{1}{2}dz \cdot X+\tfrac{1}{4}dx \cdot dz \\ \hline XZ+\tfrac{1}{2}dx \cdot Z+\tfrac{1}{2}dz \cdot X+\tfrac{1}{4}dx \cdot dz \end{array} \qquad \begin{array}{l} X-\tfrac{1}{2}dx \\ Z-\tfrac{1}{2}dz \\ \hline XZ-\tfrac{1}{2}dx \cdot Z \\ \quad -\tfrac{1}{2}dz \cdot X+\tfrac{1}{4}dx \cdot dz \\ \hline XZ-\tfrac{1}{2}dx \cdot Z-\tfrac{1}{2}dz \cdot X+\tfrac{1}{4}dx \cdot dz \end{array}$$

By subtracting these two results, he gets $dx \cdot Z + dz \cdot X$ as the differential of the rectangle XZ.

A third method. The differentials of the sides X and Z are dx and dz, and therefore the sides of the rectangle become $X + dx$ and $Z + dz$, and the rectangle itself is $XZ + dx \cdot Z + dz \cdot Z + dz \cdot dx$. By subtracting the given rectangle XZ from this result, the remainder $dx \cdot Z + dz \cdot X$ gives the differential of the rectangle XZ (the term $dz \cdot dx$ being infinitely small in comparison of either of these).

These three methods suggest a satisfactory way of presenting the steps used in obtaining the fundamental rule for differentiation to a class of beginners. This suggested method, which is used in the senior high-school calculus, first presents the derivative as the limit of a geometric ratio. Then there is obtained directly from this geometric representation the algebraic representation of the derivative as the limit of a fraction. Then from an analysis of this last form, the four steps of the fundamental rule are obtained directly.

Christian Wolf. Wolf also used a geometric method for finding the differential of a product.[12]

His supposition is that if two quantities are multiplied together as xy, then the differential of one factor multiplied by the other factor for each of the factors and the sum of these two products, which is $xdy + ydx$, will be the differential desired. That is, $d(xy) = xdy + ydx$.

In this demonstration, xy is represented by the rectangle $CABD$ whose base $AC = x$, and altitude $DC = y$. If we suppose the sides to be increased, then CA changes into $CL = x + dx$, and CD into $CE = y + dy$, and the rectangle $CABD$ changes into $CLGE$.

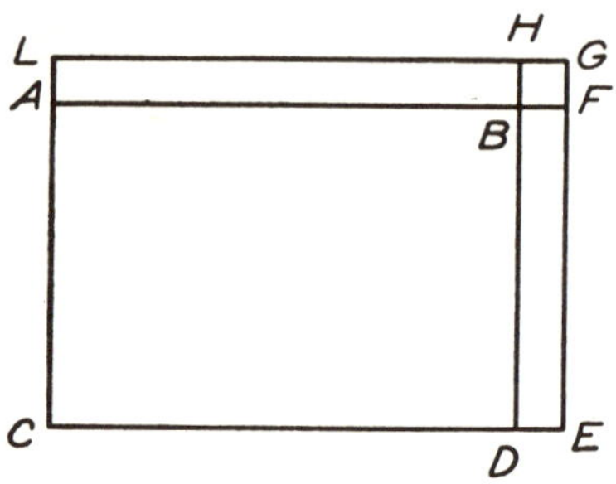

FIG. 9

Thus, the differential of xy itself is the difference between the rectangles $CABD$ and $CLGE$. Whereby, $d(xy)$ is found from $xy + ydx + xdy + dxdy - xy = ydx + xdy + dxdy$; namely, $ALHB + DBFE + BHGF$. Then in the rectangle $ALHB = ydx$, $AL = dx$ is a finite quantity; and $HGFB = dxdy$ will be its difference. But this is also the difference in the rectangle $DEFB$. Accordingly, since $HBFG = dxdy$ is the difference of each of the two rectangles $ALHB$ and $DEFB$, its relation to these two rectangles is cancelled; consequently, the difference between the rectangle $CABD$ and $CLGE$, or the differential of xy itself, is $ydx + xdy$.

Although Wolf's method is an improvement in the method for

[12] Christian Wolf, *Elementa Matheseos Universae*, Second Edition, V. I, p. 545, Magdeburg, 1730.

finding the differential of a product it was not until Cauchy gave a rigorous demonstration of his method of limits that a completely satisfactory form was obtained. Cauchy's method will be commented upon later.

Wolf extends the content of the formula $d\ (xy) = xdy + ydx$ to include the cases of the product of several variables and also of the powers of a single variable by means of a substitution method. Suppose that the expression to be differentiated is the product of three variables vxy. Let $vx = t$, then $vxy = ty$, and consequently, $d\ (vxy) = tdy + ydt$ by use of the formula, $d\ (xy) = ydx + xdy$. Then for t substitute its value vx, and we have $d\ (vxy) = vxdy + vydx + xydv$.

This method could probably be used to advantage in a calculus course in senior high-school mathematics.

Leonhard Euler. Euler was the first to introduce the symbol Δx for the difference or increment of x.[13]

M. Cousin. Cousin was a strong advocate of the method of limits. He uses two propositions from geometry as illustrations of the method of limits. One of these was first used by Bryson, and has already been mentioned. It is the proposition that states that the circumscribed and inscribed polygons approach the circle as their limit. The other is the proposition concerning the ratio of the area of the ellipse to that of the circle described upon its major axis as a diameter.[14]

Both these propositions are contained in the geometry of the senior high school, and hence are within the mental grasp of the pupils of the senior high school.

Lazare Carnot. Carnot gives a rigorous discussion on the proper use of infinitesimals and of limits. He shows that the method of infinitesimals is imperfect but satisfactory and that the method of limits is both exact and satisfactory.[15]

Augustin Louis Cauchy. Cauchy gives a rigorous proof of the method of limits as used in finding the derivative.[16]

[13] Moritz Cantor, *Geschichte der Mathematik*, V. 3, p. 725, Leipzig, 1898. This page is a part of Cantor's discussion on Euler's Differentialrechnung of 1755.

[14] M. Cousin, *Leçons de Calcul différentiel et de Calcul intégral*, V. I, pp. x–xiv, 1777.

[15] Lazare Carnot, *Reflexions sur la Métaphysique du Calcul infinitésimal*, Paris, 1797.

[16] M. Augustin Cauchy, *Leçons sur le Calcul différentiel*, p. 17, Paris, 1829.

He shows that $\Delta y = f(x + i) - f(x)$, where $\Delta x = i$, an infinitesimal. He then writes $\frac{\Delta y}{\Delta x} = \frac{f(x + i) - f(x)}{i}$ and says that as the two terms Δx and Δy approach zero as a limit, the value of this same expression will approach another limit, either positive or negative; provided, of course, a limit exists. This limit determines a value for each particular value of x, but varies also with x. Thus, for example, if $f(x) = x^m$, where m is an integer; then the expression $\frac{f(x + i) - f(x)}{i}$ becomes $\frac{(x + i)^m - (x)^m}{i}$. This will give $mx^{m-1} + \frac{m(m-1)}{1.2} x^{m-2} i + \ldots + i^{m-1}$, by means of the binomial theorem and the combining of terms and the division by the denominator i. This expression then has for its limit, as i approaches zero, the quantity mx^{m-1}, a new function of x. The same will be true in general; only, that the form of the new function which is the limit of the expression $\frac{f(x + i) - f(x)}{i}$ depends for its form upon the given function $y = f(x)$. Hence, in order to indicate this dependence, the new function is given the name "derived function" or derivative. Its symbol is $f'(x)$.

Now that the method of obtaining the derivative by the use to limits as described here has been rigorously demonstrated by Cauchy, this method may be used in future work without again proving it.

CONCLUSIONS

This historical survey furnishes some valuable suggestions as to the way the elementary calculus should be presented to a class of senior high-school pupils. There has been much use made of geometric representations. Frequently these have then been expressed in algebraic form. Then from an analysis of these algebraic forms, the working rule or the principle of the process has been obtained.

This survey also shows that mathematical intuition has played an important part in the development of the calculus. Mathematical intuition made possible methods that, although not exact, were satisfactory.

The illustrations and the discussions of the topics in this historical survey show that the introductory ideas of the elementary calculus are not too difficult for pupils in mathematics in the last year in the senior high school.

V

COMPARISON OF TEXTBOOKS ON THE ELEMENTARY CALCULUS FOR BEGINNERS AND FOR SELF-INSTRUCTION

THE STANDARD METHOD

A thorough investigation has already been made as to the nature of that part of the calculus suitable to our high schools. The conclusion is that "The work in calculus should be largely graphical and closely related to that in physics; the necessary technique should be reduced to a minimum by basing it wholly on algebraic polynomials. No formal study of analytic geometry is presupposed beyond the plotting of simple graphs." [1]

With this assumption, it would be well, in the examination of textbooks, to notice how they contribute to the standard thus set forth.

Furthermore, it is desirable to consider the method of approach to the subject which characterizes these various works. For example, does the author use the method of infinitesimals or does he favor the method of limits? Is the treatment made from the scientific viewpoint or from that of the learner's ability to grasp the subject? If the psychological method is followed, a work intended for younger pupils would naturally be genetic in character, being based upon the historical steps by which the calculus developed in the mind of the race.

The scientific viewpoint is illustrated in the older form of textbooks on high-school algebra. Many definitions are first given, many abstract ideas are explained in detail, and a number of topics of little value to the learner such as the extraction of the roots of polynomials are treated at length. The scientific method is the proper method for mature minds with sufficient preparatory training in mathematics but, as experience in teaching algebra has shown, it is not the method best suited to the minds of younger pupils.

[1] *Elective Courses in Mathematics*, fourth revision, a preliminary report by The National Committee on Mathematical Requirements, March, 1921.

TEXTBOOKS REVIEWED

AMERICAN. There are almost no American textbooks of this type.

Elementary Analysis, by W. A. Granville and P. F. Smith (Boston, 1910). The first part of the book is devoted to analytic geometry and the second part to the elementary calculus. The object of the book is to prepare a simple and direct exposition of those portions of mathematics beyond trigonometry which are of importance to students of natural science. There is an intentional avoidance of anticipatory difficulties. Processes which are natural are introduced without explanation. Much use is made of graphical representation.

Calculus with Applications, an Introduction to the Mathematical Treatment of Science, by Ellen Hayes (Boston, 1900). Its aim is twofold: for purposes of culture and for the purpose of learning, in as simple a way as possible, what the calculus is and what it is for. The book is a reading lesson in mathematics and the examples are taken from mechanics and astronomy. Relationship is expressed as "a quantitive change in the cause which is accompanied by a quantitative change in the effect." The term function is applied to the quantity which necessarily changes because of a change in a variable with which it is connected. The chief object of the book is to show how to use the operations of the calculus.

Elementary Calculus, by William F. Osgood (New York, 1921). This book, although intended for college use, is included in this review because it is probably the latest textbook on this subject. "The object of this book is to present the elements of the Differential Calculus in a form easily accessible for the undergraduate." Professor Osgood emphasizes the use of graphical methods.

ENGLISH. There are a number of English textbooks on the calculus for beginners in which the authors have simplified the treatment of the subject.

The Calculus for Beginners, by W. M. Baker (London, 1919), "The differentiation and integration of the simpler standard forms are applied as early as possible to the determination of maxima and minima, of the areas and lengths of curves, of volumes of revolution, to the solution of problems in mechanics and physics and to the expansion of simple trigonometric functions." The author treats the method of limits at considerable length. Graphic illustrations are frequent. The book contains 750 examples largely of an applied character.

Differential Calculus for Beginners, together with a second book, *Integral Calculus for Beginners*, by Alfred Lodge (London, 1913). The object of the differential calculus is to provide an easy introduction to the subject for those students who have to use it in their practical work, by making them familiar with its ideas and methods within a limited range. "Stress is laid at the outset on the graphic representation of functions, as by this means a vividness and reality is given to the processes of the calculus which is not easily obtained without such ocular assistance. In this way the theory of maxima and minima becomes very simple and interesting." The author introduces the differential before the derivative and gives an excellent discussion of the subject. He introduces the differential coefficient or derivative by simply stating that it is the ratio dy / dx when $y = f(x)$. His integral calculus is written for students of physics and mechanics and contains many applied problems.

An Introduction to the Differential and Integral Calculus for the Use of Engineering and Technical Students, by John Stoney (London: no year).[2] The practical side of the subject is dealt with. "When two quantities are so connected that a definite value given to the one produces a corresponding definite value in the other, each is said to be a function of the other." The author introduces the derivative as a rate. Throughout the book the work of the calculus is applied without any attempt at an exhaustive and rigorous treatment of the theory. However, he frequently introduces abstracts of such a treatment. This makes the work lack unity. The author includes such subjects as partial differentiation and hyperbolic functions. The pupils would require a knowledge of physics and engineering.

The Elements of the Differential Calculus, by W. S. B. Woolhouse (London, 1852). Woolhouse uses the method of infinitesimals, and says that this method is just as rigorous as that of limits. "In comparing the relative values of any two infinitesimals, the rejection of terms involving infinitesimals of higher orders is, in effect, precisely the same as that of proceding to the ultimate ratio of the infinitesimal quantities by the method of limits and such rejection

[2] The Library date is 1920.

of values may be said to be the operation of cropping down the quantities to their ultimate or limiting relative positions."

Easy Lessons in the Differential Calculus; Indicating from the outset the utility of the processes called differentiation and integration, by Richard A. Proctor (London, 1887). Proctor approaches the subject through the laws of falling bodies, and spends several pages on this discussion. He treats the integral calculus simply as a part of the differential calculus, and considers "the differential calculus as the science which deals with the rate at which variable quantities increase or diminish."

Calculus Made Easy, by S. P. Thompson (London, 1919). The book is written in very simple and interesting language and shows how easy some of the apparently difficult things in calculus may be made. The author first explains in unscientific language the meaning of the symbols for differentiation and integration. He discusses each new topic at considerable length, and often explains the difficulties in near-slang language. He makes considerable use of graphical representation. "The aim of this book is to enable beginners to learn its language, to acquire familiarity with its endearing simplicities, and to grasp its powerful methods of solving problems, without being compelled to toil through the intricate out-of-the-way (and mostly irrelevant) mathematical gymnastics so dear to the unpractical mathematician."

FRENCH. There are no separate French textbooks on the elementary calculus for beginners since this kind of work is included in their courses in algebra.

Précis d'Algèbre, by Carlo Bourlet, Paris, 1909. This book includes the work on the calculus as prescribed in *Plan d'Études et Programmes d'Enseignement dans les Lycées*, 1907–1908. The outline of this work is given in Section I of this dissertation under the *Lycées* (France).

Algèbre, Second Cycle, by Émile Borel, 3rd ed., (Paris, 1905). This book, also, was written to meet the requirements as set down by *Programmes Officiels de 1905*. This work for the *Classe de Seconde* (Sections C and D) in Algèbre, so far as it relates to the simple calculus, has already been indicated in Section I of this dissertation under the *Lycées* (France).

Cours d'Algèbre Élémentaire Conforme aux Derniers Programmes de l'Enseignement Secondaire (1902), by F. G.-M. (Paris, 1909). As the title indicates, this book also gives the subject-matter as required by the *Programmes Officiels*. However, the subject-matter is more comprehensive than in the two preceding books, and there are also a greater number of applications of the derivative. The graph work also is discussed at greater length than in the preceding books.

Cours de Mathématique à l'Usage des Candidats à l'École Polytechnique et Autres, par Charles De Comberousse, Tome Troisième (Paris, 1887). The scientific method is used. The author gives a good classification of simpler functions into algebraic and transcendental functions and their subdivisions.

GERMAN. There are a number of German textbooks on the calculus written for beginners and for self-instruction.

Elementarbuch der Differential- und Integralrechnung mit Zahlreichen Anwendungen aus der Analysis, Geometrie, Mechanik, und Physik, für höhere Lehreanstalten und den Selbstunterricht, von Dr. Alfred Donaldt (Leipzig, 1901). The first part consists of an intuitive introduction into the calculus by means of simple differentiation and integration. Then the scientific method is used in separate treatments of differentiation and integration.

Elemente der Differential und Integral Rechnung, by Ludwig Tesar (Leipzig, 1906). This book is probably within the intellectual capacity of our senior high-school pupils. Asymptotic curves are used to prepare the learner's mind for the work in limits. He introduces the differential coefficient by means of a time-speed table. Integration is introduced as a summation process.[3]

Die Anfangsgründe der Differentialrechnung und Integralrechnung, by Dr. Richard Schroeder (Leipzig, 1905). This book is intended for self-instruction and for use in the higher schools. However, it does not differ materially from the traditional calculus texts for use in colleges. It was written to meet the needs of the new course of study to be introduced into the Prussian secondary schools. The book would be unusable in our senior high school.[3]

CONCLUSIONS

This review of textbooks shows the tendency to put the study of the elementary calculus earlier in the mathematics course. It also shows that textbooks on the elementary calculus are being written

[3] These two books were reviewed by Mr. M. A. Norgaard.

now from the standpoint of the learner's ability to grasp the subject. The English textbooks emphasize graphical representation and the abundant use of material from physics and mechanics. Their graphical representation necessarily requires the use of the algebraic polynomial. Such books are the most hopeful of any we have in teaching the calculus to younger pupils.

A study of these works would seem to show that a course in the elementary calculus in the senior high school should have its theory based upon algebraic polynomials with the explanations given by means of graphical representations. The work should be closely allied to physics and mechanics by the use of a sufficient number of suitable problems from these fields to give reality to the subject and an insight into the wide use of the calculus in applied fields.[4]

[4] This is the method that is followed in the course of which a synopsis is given in Section VII of this study.

VI

THE TREND OF AMERICAN EDUCATION IN GENERAL

THE INCREASED LENGTH OF SCHOOL ATTENDANCE

Not many years ago the educational scheme for our public schools was extremely limited. A few years of a few months each was the extent of the intellectual training of a pupil, and this was looked upon as sufficient for the conditions of those times. As conditions changed, so also did the scheme of education, but less rapidly. The reason for this is that our educational system, like the church in this respect, is among our more conservative institutions.

Before the beginning of our present attempt at a readjustment in our educational system of which the initial wave is the junior high-school movement, the type child attended school eight years of nine months each. There were some children who began in the kindergarten and completed a four-year high-school course. The educational scheme for these children covered from twelve to fourteen years. To-day there are indications that the period of compulsory education of a certain sort should be extended to the age of eighteen years, such as part-time education for young people who leave school at the age of fourteen to become wage-earners. Also, the bright pupil who is without the necessary financial means may be able in the future to extend his number of years in study by the aid of a government subsidy.[1] There is an inclination, too, on the part of the Labor Party to favor an extension of the age limit of compulsory education in order that its members may enjoy greater security against the encroachments of child labor.[2]

It is also true that many young people continue their studies for many years after they have entered the work-a-day world. The discussion thus far shows that there is a marked tendency to render indistinct the dividing line between the period of formal education as represented by our public school system and the time when the individual ceases educational activity. This raises the question,

[1] Notes on lectures by Professor S. S. Colvin, Teachers College, summer, 1918.
[2] Notes on lectures by Professor D. E. Smith, Teachers College, winter, 1920.

is there a limit as to the number of years one should indulge in study, especially in the study pertaining to a special field?

THE CHANGED POINT OF VIEW IN EDUCATION

We hear this question, is not the time spent in our present scheme of education already too long? Or, does it not take too many years out of the lives of the children in preparation for life? This latter question must be answered in the affirmative, if in thinking of education as a preparation for life we consider that the pupil should be more or less isolated from the ordinary interests of life during this period and that he should be imbibing theory which in later years he shall put into practice in meeting life's situations. There is too much indefiniteness and uncertainty of attainment of the desired end in thus looking at education as a preparation for life itself.

However, the scheme of education resulting from our present attempt at readjustment in our educational system considers education not merely as a preparation for life but as life itself. By the statement that education is life itself, among other things, is meant that the pupil is forming habits that he will desire to maintain and strengthen throughout his lifetime, that the methods of obtaining knowledge that he is acquiring are the kind of methods that he will use for this purpose throughout his lifetime, that the kind of thinking in which he is being trained is the kind of thinking that will serve him best in meeting other situations in life, that the ideals he is forming will be the nucleus for the development of those more mature standards which will serve as the stabilizers and guides of his life, and that the attitudes that he is being led to assume are such that they will need no radical change but merely accentuation to make him an acceptable member of human society. Of course, this new way of viewing education carries over much that has always been in education, but this changed point of view is highly important in that it makes for a better selection of the material that should be included in our school work and of the methods in general in secondary education. This changed point of view means that educational activity is a constituent, not an incidental, part of life itself, and that among the more intelligent individuals, and especially the leaders of human society, educational activity never ceases.

THE VIEWS OF OTHERS

Professor W. H. Kilpatrick considers that the typical unit of educational procedure should be "wholehearted purposeful activity proceeding in a social environment, or more briefly, in the unit element of such activity, the hearty, purposeful act." [3]

Professor David Snedden says, "In a sense any concrete job undertaken in a vocational school where the realization of valuable results in the product constitutes an important end, might be called a 'project,' but to be an 'educational project' such a job . . . must be of such a nature as to offer large opportunity, not only for the acquisition of new skill and experience in practical manipulation, but also for applications of old and learning new 'related knowledge', art, science, mathematics, administration, hygiene, social science, etc." [3]

Professor E. L. Thorndike expresses a more general view of education. He says, "The arts and sciences serve human welfare by helping man to change the world, including man himself, for the better. The word education refers especially to those elements of science and art which are concerned with changes in man himself." [4]

One never gets too old to study provided he has had sufficient earlier training in habits of mental activity. That an individual who continues to develop and strengthen his mental habits until he has reached the age of twenty-five can, if he wishes to, assimilate new thought for the remainder of his lifetime, is a frequent interpretation of a discussion on habit by a world-famous psychologist.[5]

The Need of More Trained Mathematical Minds and the Source of Supply. Since America has become the greatest manufacturing and engineering nation in the world, her need of more men with a wider knowledge of mathematics is rapidly making itself apparent. Hence, the elementary calculus would be the logical subject to supply the want of more vigorous and profitable mental food for the bright pupils of our senior high school who are mathematically inclined. That our bright pupils are in need of more vigorous mental food has been confirmed by investigation. If we divide pupils into three classes, bright, ordinary, and dull, and compare their educa-

[3] Notes on lectures by Professor William C. Bagley, Teachers College, summer, 1920.

[4] E. L. Thorndike, *Educational Psychology*, V. I, p. 1, New York, 1919.

[5] William James, *Principles of Psychology*, V. I, pp. 121, 122.

tional attainments with their general intelligence ability, it will be found that the last two classes have reached a standard for their educational attainments that is as high as their general intelligence ability will warrant, but that the bright pupils fall below in their standard of educational attainments as compared with their general intelligence ability. This waste of fertile mental resources is not only an individual loss but also a national one, and should be promptly eliminated.[6]

The following quotation gives the modern point of view in education, "We are attempting to use every bit of native ability of the pupil and by so doing eliminate the waste of intelligence that is now going on in our schools to-day. The wasted intelligence is much greater in amount than most of us have realized. Ever since the introduction of mental and educational tests we have all been aware of this wastage, but we have not been able to measure it accurately. The combination of mental and educational tests will enable us to do so. It will show us just where—upon what school or class or pupil—pressure must be brought to bear in order to eliminate the waste and make the school, class, or pupil work up to the extent of the native capacity.

"As a matter of fact we shall see, what we long suspected and what we have only recently consciously realized, that the greatest amount of waste exists among the brighter pupils in a class or among the better schools in a school system. It is as a rule the more intelligent pupils that are working below capacity, even although they are keeping well up to the average level of the group. We have been pushing and cramming the duller children while the brighter ones have been allowed to loaf. The bright child is the most retarded child in our schools. The dull child is the most accelerated. The bright child is the laziest child, and the dull child is the most industrious. We are now ready by means of combined mental and educational tests to equalize the pressure, and stimulate both the bright and the dull so that they may work up to their respective capacities. If we can accomplish this, we shall have happier and better children. Lack of adequate stimulus undoubtedly leads to much disciplinary trouble in our schools. Habits of mental laziness acquired in school often persist through life, and there are undoubtedly many adults at the present time,

[6] Notes on a lecture by Dr. Rudolph Pintner, summer, 1920.

who have splendid native ability and who do not know it because the school has taught them to be satisfied with a mediocre type of accomplishment."[7]

CONCLUSIONS

It has been shown that the bright pupils in our high school need more vigorous food. A course in the elementary calculus will satisfy this need for those who are mathematically inclined.

It has been shown that many individuals study along certain lines the greater part of their lifetime. The person who may study along the line of applied mathematics will need a knowledge of the two operations, differentiation and integration, in order to understand some of the most valuable mathematical contributions. Such a person would have received in a course in the elementary calculus in the senior high school sufficient instruction in differentiation and integration to continue further his study of the calculus without the aid of a teacher.

[7] Rudolph Pintner and Helen Marshall, "A Combined Mental-Educational Survey," *Journal of Educational Psychology*, Jan., 1921, p. 32.

VII

SUGGESTIONS FOR A MODERN PRESENTATION OF THE ELEMENTARY CALCULUS

THE AIM OF THESE SUGGESTIONS

These suggestions are intended to show how to simplify the subject-matter of the elementary calculus and its presentation, how to make the topics definite, and how to lead the pupils to see the theory as a related whole. A feature of this high-school course in the elementary calculus which tends to make the work definite is that the general purpose of the course harmonizes with the general purposes of education. One of these general purposes of education is to impart to the coming generation the knowledge which has become the heritage of the race.[1] It has been shown that the calculus is a part of this heritage since it is the product of the minds of many men throughout many years. In the selection of the material for this course and in the manner of the presentation of this material, the particular aim of the course has been kept constantly in mind.[2] This is a feature that contributes materially to the definiteness of the work. The unity of the theory has been made to stand out more clearly by the use of the same kind of material in nearly all of the explanations. The presentation of the subject-matter has been simplified by making use of material that lies within the experience of the pupil. This experience may be real or imaginary.

THE DIVISIONS OF THE DISCUSSION

The first part of these suggestions deals with the plan of the course, the aim, and the method of presentation. The second part deals with the particular topics included in the course. This second part is written in the form of a synopsis of a text-book on the elementary calculus suitable for use in our American high schools.

[1] Notes on lectures on the Technique of Teaching, by Professor W. C. Bagley, summer, 1920.

[2] See statement of the particular aim a few paragraphs later under the heading "The Aim of the Course," p. 48.

First Part

GENERAL FEATURES INFLUENCING THE PLAN OF THE COURSE

The Viewpoint. There are a number of textbooks on the calculus for beginners and for self-instruction. Many of these textbooks were written by mathematicians for use in schools abroad. With probably one or two exceptions they are not satisfactory for use in our senior high school because our schools differ from those abroad both in the character of the pupils and in the nature of the organization. The pupils in our senior high school come from a greater variety of walks in life than is the case in Europe, and although they probably possess greater native ability this ability is less highly trained. Also, in our high schools there is little offered in the curriculum that leads definitely to the strong technical courses of more advanced work as is the case in schools abroad. Hence it is necessary that we consider the plan of a course in the elementary calculus from the point of view of conditions in our secondary schools and especially as these relate to the mental needs and capacities of the pupil.

Important points in estimating the status of a course or subject of study in our secondary schools are use, appreciation, mental discipline, interest, and degree of difficulty. These features will now be discussed.

Use. Without the calculus much of the work in engineering would be impossible, a large part of the work in science would be crippled, and a large field of development in pure mathematics would be eliminated. However, an understanding of the work in any one of these fields must go hand in hand with a knowledge of the calculus if a person wishes to apply the latter in such a field. Hence the skill and knowledge imparted in a calculus course in the senior high school does not equip the pupil to make any worth-while use of his knowledge of the subject in applying it to real conditions within his experience at the time, but it does enable him to use the calculus that he will find in the field in which he intends to specialize later. This preliminary course also tends to build up that hierarchy of habits which is necessary to a satisfactory understanding of the more advanced work in the calculus.

In the realm of pure mathematics the vast fields of function theory, integral equations, and differential equations make continual use of the calculus. Indeed, it would be a great advantage to the investigator in these fields if his calculus had grown up with him in the way that his geometry, algebra, and trigonometry have done.

The calculus unifies the related mathematics that has preceded it since it makes use of arithmetic, geometric, algebraic, and trigonometric ideas indiscriminately as occasion demands. It also shows the pupils the use of many of the things that they have studied in the preceding branches of mathematics and thus serves especially well as a practical review of the parts of the preceding mathematics that are worth remembering. The calculus is also found useful to the average well-informed citizen in his general reading of scientific articles of a popular nature.[3]

Appreciation. A course in the elementary calculus in the high school will add to the pupil's appreciation of mathematics since it will open up to him vistas of the wide application of mathematics. This course also acquaints the pupil with two new operations, differentiation and integration. These two tools are so powerful that the pupil cannot help but marvel at the far-reaching influences of the principles of mathematics.

The appreciation of a subject is a point worthy of greater consideration than may at first be apparent. Since the appreciation of the subject contributes toward the cultural development of the learner, it also contributes toward the general development of culture in our country as a whole. This is an important matter, since, because we have become the richest country in the world, it should not only be our desire but it is also our obligation to strive toward a high standard of culture.

Disciplinary Value. Educational psychology has shown the fallacy of the view "that one of the chief values of education is to be found in the cultivation of such mythical powers of the mind as reason, attention, imagination, discrimination, and the like."[4] That is, even if it is conceded that a subject has disciplinary value, this reason alone would not be sufficient for including it in the curriculum. However, if the subject has disciplinary value of a

[3] The Amount of Mathematics the General Reader should Know as Found by Scanning the Articles in a Standard Encyclopedia: a class investigation conducted under the direction of Professor D. E. Smith, winter, 1920.

[4] S. S. Colvin, *An Introduction to High School Teaching*, New York, 1918.

particular nature that is highly desirable, then its disciplinary value would be a contributory reason for including the subject in the curriculum.

In the case of the elementary calculus, an outstanding feature is the specific training that the learner receives in the function concept. There is no attempt made to develop the learner's ability to recognize the underlying function relations of the conditions under consideration by means of a mere presentation of the theory of the calculus given in abstract form. This ability is developed largely, to a greater or less extent, unconsciously in the learner's efforts to apply the theory to the material taken from the fields of science, engineering, physics, mechanics, aviation, wireless telegraphy, and the like, and to the simpler material given earlier in the course which has been suitably modified so that the pupil will not go far astray in the learning process.

In brief, since the calculus is the mathematics of nature and treats of small changes in related conditions, it develops the kind of thought that everyone makes use of in considering the problems that come up in our daily lives. In fact, one college professor of mathematics has stated that the elementary calculus consists of ninety-nine per cent common sense and one per cent calculus.[5]

Interest. There should be little attempt to arouse interest artificially in the elementary calculus by means of far-fetched problems from specialized fields. Indeed there need be no resort to artificial means to stimulate the learner to an interest in the subject, since he has chosen the elective courses in the mathematics of the senior high school because he likes mathematics or because he intends to follow lines in which he will have further need of the science. Nevertheless, in the exercises of the course in the elementary calculus there are included many applied problems that will contribute toward stimulating an interest in the subject.

Degree of Difficulty. By degree of difficulty is meant, is the subject within the mental grasp of the pupils for whom it is intended? Since the elementary calculus is taught in many schools abroad to pupils of approximately the same age as our pupils in the twelfth school year, or the last year of the senior high school, it surely is not too difficult for our selected pupils whose tastes have led them to

[5] Notes on lectures by Doctor H. F. Stecker, The Pennsylvania State College, winter, 1910.

elect to study the subject. In fact, this introductory course in the elementary calculus does not contain work as difficult as that required to solve many of the puzzle problems and problems on obsolete subjects in the arithmetic of bygone days. Furthermore, a historical survey of the development of the calculus shows that the more elementary topics lie easily within the grasp of pupils in the last year of the senior high school.

Hence, the course in the elementary calculus is useful, it generates an appreciation of mathematics, it develops the kind of thought that can be applied to other situations in life, it is interesting to the pupils who elect mathematics, and it is not too difficult for such pupils.

THE AIM OF THE COURSE IN ELEMENTARY CALCULUS IN THE SENIOR HIGH-SCHOOL MATHEMATICS

The aim of the course is twofold: to give to the pupils the beginning of the mathematical language they will have to use if they continue further their studies in mathematics; and to give to pupils such an introduction into the subject as will enable them to pursue its study further without the aid of a teacher.

In working to attain this aim there will be built up, as has already been stated, that hierarchy of habits which is necessary to a satisfactory understanding of the more advanced calculus. Also, this course in the elementary calculus will be found to contribute to the satisfaction of the innate desire of such pupils who display a love for mathematics.

THE METHOD OF PRESENTATION

The method of presentation is molded largely according to the suggestions derived from a historical survey of the development of the calculus in the mind of the race. This development was simple and natural. Since the development of the individual mind is to some extent analogous to the development of the mind of the race, many of these primitive methods will be found to be well worth following in teaching the elementary calculus to high-school pupils.

The suggestions derived from this historical survey emphasize: first, the geometric representation; second, the algebraic representation; and third, the obtaining of the law or rule by an analysis of these two representations. Furthermore, in the historical development of such topics as the derivative it will be seen that there are overlapping phases of intuition, demonstration, and application. It will be found that frequently there were sound deductions, but very often the rigorous proof was not given until many years later—and so it may properly be in presenting the subject to pupils at the present time.

THE MATERIAL

The Use of Graphs. The simplest vehicle for the presentation of the theory to beginners is that of graphs. This simplicity is increased in the method suggested in this study by the use of the same specific curve in the explanations of nearly all the topics of the theory. In this way the pupils will have more mental energy left free for concentration upon the new ideas which the calculus presents. These graphs are already familiar to the pupils, they are easily understood, and they are adaptable, economical, and free from the time-consuming, irritating explanations necessitated by technical material taken from the applied fields. A certain amount of material from these latter fields will, however, be included in the exercises, particularly such as will be easily understood by the pupils. To use throughout the explanations of the theory one particular kind of material, such as graphs, that is readily understood by the learner, not only adds to the simplicity of the treatment but also aids the pupils in perceiving the unity that exists throughout the theory.

The material of the course, although given as a connected whole, has three divisions. The theory is treated in such a way that it may be separated throughout from the exercises and be made a part of the course by itself. The first part of the exercises under each topic is not too difficult and is selected in such a way as to illustrate the ideas that have just been explained and to give the pupils practice in the processes involved. The second part of the exercises under each topic includes carefully selected interesting applied problems. It is left to the option of the teacher to decide which of these latter problems the pupils should attempt.

In the topics that follow in this section the theory of a course in the elementary calculus in the high school will be given in fairly complete form, but only an illustrative example or two will be given of each of the two kinds of exercises just described.

Second Part

THE METHOD OF APPROACH TO THE SUBJECT

The method of approach to the elementary calculus and to each of the topics discussed is left to the judgment of the teacher. A teacher who has had training in physics would probably find it desirable to introduce the subject of the calculus through some such problem as that of a falling body. Another teacher who has not had much training in physics would probably find that a talk upon the uses of the calculus would be sufficient to arouse immediate interest in the subject. Many illustrations of the uses of the calculus may be found in textbooks on the subject and in technical journals. Very likely a single illustration, such as the tinsmith's problem (given later under the heading, A Maximum Illustrated by a Practical Problem), will give the class sufficient incentive to start on its study for obtaining the rule for finding the derivative. Beginning with the topic, The Function Idea, all the topics are developed as a part of the process for obtaining the fundamental rule for finding the derivative of a function. It should be constantly kept in mind that the deriving of this fundamental rule is the main objective of this part of the theory.

In the synopsis of the course in the elementary calculus in the senior high-school mathematics, the theory is based upon algebraic polynomials with the explanations given by means of graphical representations. Throughout these suggestions, illustrative examples are given which show how the work may be closely allied to physics and mechanics. The author believes that this is the preferable method since it is the shorter road into the mathematical insight of the nature of the calculus, since mathematical teachers in general (unless they have had a strong preparation in physics) will find this the more convenient method, and since also the pupils (unless they too have had a good preparation in physics) will be able to grasp this method more readily. Furthermore, it is very probable that the most difficult thing in the making of a high-school course in the calculus is the simplification of the mathematics of the calculus in such a way that the deductions made and the statements given will be sound and in agreement with results as obtained by the rigorous methods of the advanced work in pure mathematics. The author has been exceptionally careful to keep this difficulty in mind and is confident that the method he has herewith outlined will form a reliable standard for the teacher of secondary mathematics to follow if he wishes to keep the mathematics of the course in elementary calculus correct.

What is the Calculus? It is likely that the student has heard mention made of the calculus in his previous courses in mathematics. No attempt will be made here to give a complete idea of this branch of mathematics. Only an illustration will be given to indicate in general the nature of the subject.

Suppose that we had a steel bar n inches in length at a given temperature. If we increase the temperature, we will increase the length of the bar slightly. We say that the length of the bar has taken on an increment. If we decrease the temperature, the length of the bar will be decreased slightly. This is also an increment but in the negative direction. Hence, we say that the bar varies in length as the temperature varies. In this respect the length depends upon the temperature; that is, the length of the bar is a function of the temperature.

There are many such related changes in nature. However, these related changes need not be according to natural laws only, but may be due to other causes. The calculus is based upon a consideration of the variations of such function relations.

A Reason for Studying the Calculus. Thus far the student has learned several valuable operations for calculation; *i. e.*, addition, subtraction, multiplication, division, involution, evolution, and logarithms. The calculus will add two more operations to these, namely, differentiation and integration. Without these last two operations, much of the mathematics necessary in engineering and in the sciences would be impossible. Because of the many formulas, the actual calculations in the calculus are not so very difficult, but in order to master the mechanics of the subject it is necessary that the introductory work should be thoroughly understood.

The Previous Work in Graphs. Before undertaking the study of this proposed course in the elementary calculus, the class is supposed to have completed the senior high-school courses in demonstrative geometry, algebra, and trigonometry, or their equivalent. If their work in graphs did not include the finding of maxima and minima values, this topic should be given before the work in this proposed course is taken up.

An illustrative example. A patriotic boy who wished to do

his part during the world war made up his mind to have a war garden. There was a vacant lot nearby partly covered with rubbish that would give him plenty of space for the garden. A neighbor donated a piece of chicken-wire. The boy measured the wire and found it to be 48 ft. long. The question was this, how wide and how long should the garden be, if rectangular in shape, in order that the given wire might enclose the greatest space possible? The boy tried different dimensions and used a graph to help him to see the results in a better form and to serve as a check on the correctness of his work.

Solution. The width plus the length would be equal to ½ of 48, or 24. The boy then made the following table:

Width	Length	Area
6	18	108
8	16	128
10	14	140
12	12	144
14	10	140
11	13	143
13	11	143

Then he made the graph from this table. Both his calculation and his graph showed him that apparently the dimensions 12 by 12 gave the greatest area.

The method of the elementary calculus for this example is as follows. Let x be the width, then the length is $24 - x$, and $x(24 - x)$ or $24x - x^2$ represents the area. The derivative of this expression is $24 - 2x$. Set the derivative equal to zero and solve the resulting expression for x. This gives 12 as the width. Hence the dimensions 12 by 12 give the greatest rectangular area that can be enclosed by the given wire.

This latter method in which the derivative is used is an exact method and, for more difficult problems, a more convenient method for solving problems of maxima and minima. (See tinsmith's problem, p. 69.) A number of additional problems of this kind would arouse in the class a desire to know how to use the derivative and create a motive for studying the derivative. It should then be pointed out that the development of the rule for finding the derivative is the first important matter in the beginning work in the proposed course in the elementary calculus. After the rule for finding the derivative has been developed, the number of applications is greatly extended. The topics leading up to the rule for finding the derivative of an expression are the function idea, independent and dependent variables, increment, function notation, and the two steps in the explanations of the derivative. These will now be discussed in the order as here given.

THE FUNCTION IDEA

The Previous Training of the Pupils in the Function Idea. Although the definition of a function may not have been given in the preceding mathematics or a symbol given for its representation, nevertheless the meaning of a function has been developed to a considerable degree.[6] The idea of functional relationship is contained in much of the preceding mathematics, and the pupils are prepared for further instruction in the function idea even if it had not been explicitly commented upon heretofore.

Illustration of the Function Idea. The pupils plot the specific curve $y = x^2 - 4x + 5$. This recalls to the pupils the ideas they have learned about graphs, and brings them up to a point where their minds are prepared to receive the new ideas.

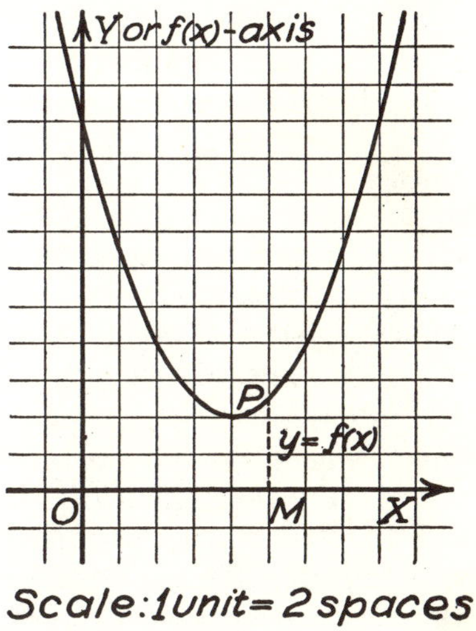

FIG. 10

Then, guided by the teacher, they select some suitable point P on the curve and draw the ordinate PM. It is seen that the value of the expression $y = x^2 - 4x + 5$ at the point P is the length of the ordinate PM. This is an important relation that is used later in the explanation of the derivative.

The expression $y = x^2 - 4x + 5$ may be read "y is an expression containing x." However, in the language of the calculus, instead of

[6] Secondary School Circular, United States Bureau of Education, No. 5, February, 1920. The ideas on the function concept contained in this circular are expanded upon and it is shown in detail how they may be applied in the classroom in the treatment of specific topics, in a report by The National Committee on Mathematical Requirements, February, 1921.

saying an expression containing x, it is read "y is a function of x." The symbol for a function of x is $f(x)$ and is read "function of x." Then in the language of the calculus $y = x^2 - 4x + 5$ would be written $f(x) = x^2 - 4x + 5$. This is read "a function of x is $x^2 - 4x + 5$." Hence here y and $f(x)$ stand for the same thing, that is, y is identical with $f(x)$. Since y represents the length of the ordinate at some point P, then the length of the ordinate may also be represented by $f(x)$.

The reason that y in the expression $y = x^2 - 4x + 5$ is called a function of x is because of the relationship it bears to x. In plotting the curve, the values of y were obtained by giving x different values. If a small change is made in x in this expression, then there will be a corresponding change in the value of y. There are many illustrations of a relationship like this between two quantities.

A simple illustration is the relation between the area of a circle and the radius of the circle. The area depends upon the value of the radius, that is, the area is a function of the radius. In symbols this may be represented thus: $A = f(R)$. The pupils should give many other illustrations of this kind.

A concrete illustration. Suppose that we wish to make a rectangular garden. The vegetables we wish to grow are to be planted at a certain number per square yard. Suppose that the length of the garden is fixed by the length of the lot but that the width of the garden will depend upon our wishes. Represent the length in yards by L and the width by x. Then the number of square yards in the garden may be represented by Lx. Since L or the length is fixed, we can change the area only by changing the width. A small change in the width will produce a corresponding small change in the area. Hence the area depends for its value upon the width, and we say that the area is a function of the width. In symbols this may be represented thus: $A = f(x)$.

Examples. Since the number of rings in a cross-section of a tree depends upon the age of the tree, we may say that this number of rings is a function of the age of the tree. If R represents the number of rings and A the age of the tree, then $R = f(A)$.

The height of the mercury column in a thermometer varies as the temperature varies. Hence the height is a function of the temperature. We may represent it in this way, $H = f(T)$.

The quantity of current generated by a dynamo is a function of the rate at which the lines of magnetic flux are being cut. If Q represents the quantity and M the number of lines of magnetic flux cut per second, then $Q = f(M)$.

INDEPENDENT AND DEPENDENT VARIABLES. DEFINITION OF A FUNCTION

In the preceding topic it has been pointed out that if in the expression $y = x^2 - 4x + 5$, we give a small increase to x then there will be a corresponding change in the value of y. Here y depends for its value upon the value that is given x, hence y is called the dependent variable. Since x may be given values at will, it is called the independent variable. Everything is now in readiness for the following definition of a function. A quantity called the dependent variable is a function of another quantity called the independent variable, if for any slight change made in the independent variable there is a corresponding change in the dependent variable.

Example. Suppose that a man walks n miles an hour. Then the distance, d, that he walks in t hours may be represented by nt. Here the distance is a function of the time. This may be represented thus: $d = f(t)$. Here t is the independent variable and d is the dependent variable. Point out the independent and dependent variables in the examples under the preceding sub-topic, a concrete illustration. Many additional simple illustrations should be given by the pupils.

INCREMENT

If in the expression $y = x^2 - 4x + 5$, x is given a slight increase, then there will be a corresponding change in the value of y. This small increase in a variable is called an increment. The term increment may also stand for a small decrease in the variable. The international symbol for the increment of x is Δx. [7] Then the increment of the value of the expression would be represented by Δy since here y stands for the value of the expression.

In the graph of $y = x^2 - 4x + 5$, some point P was selected. Suppose that the abscissa of the point P is x and that its ordinate is y. Then the coördinates of the point P are x and y. This may be represented in this way, $P \equiv (x, y)$. Now suppose that x is given a

[7] Since the differential triangle results from an increment in x, this symbol for the increment may be associated with the triangle in that way.

small increase in value which is represented by Δx. Then there will be a corresponding change in the value of y which may be represented by Δy. The sum of x plus its increment is represented as $x + \Delta x$, and the sum of y plus its increment is represented as $y + \Delta y$. These sums are respectively the abscissa and the ordinate of some other point which may be designated as P'. (See the figure under the following topic.) Then the coördinates of P' are $(x + \Delta x, y + \Delta y)$. This relation should be noticed carefully because it is used in the explanation of the derivative.

Example. Suppose that a man walks n miles an hour. Then the distance, d, that he walks in t hours may be represented by nt. Here the distance is a function of the time and may be represented as $d = f(t)$, or it may be represented fully as $d = nt$. How should this last expression be altered to show that the time has been increased slightly? Then t would have added to it an increment Δt and d would have added to it an increment Δd, and the expression may be written as $d + \Delta d = n(t + \Delta t)$.

Suppose that a railroad company charges a cents per mile traveled. Then the cost of a fare depends upon the distance traveled. Let d represent the distance and c the cost, then $c = f(d)$. Now if d is given an increment, how may this function relation be expressed? It may be expressed either as $c + \Delta c = f(d + \Delta d)$, or as $c + \Delta c = a(d + \Delta d)$.

It has been proved by astronomers by means of very accurate observations with powerful instruments that the two pointers of the north star are moving very slowly in opposite directions.[8] If D represents the distance between them at this instant and T represents an instant of time several thousand years hence, write the expression representing the distance between these two stars at the latter instant. It may be represented as $D + \Delta D$, since the increase in distance is so very slight. The pupils should give additional simple illustrations of this kind.

FUNCTION NOTATION

Numerical Illustration. In the graph of $f(x) = x^2 - 4x + 5$, if x is given a particular value, then the corresponding value of $f(x)$ is the length of the ordinate of the point whose abscissa is this par-

[8] Seymour Eaton, *How to Locate the Stars*, New York, 1891.

ticular value of x. Suppose that the abscissa of a certain point is 2.5. Then wherever x occurs in the expression, substitute the particular value 2.5 that has been given to x. The right-hand part of the expression becomes $(2.5)^2 - 4(2.5) + 5$. This is a function of x when x has been given the particular value 2.5. It is represented in this way,

$$f(2.5) = (2.5)^2 - 4(2.5) + 5.$$

Then $f(2.5) = 1.25$, the result obtained by simplifying the expression. Then 1.25 is the ordinate of the point whose abscissa is 2.5, since $f(2.5)$ represents the length of this ordinate.

Example. Suppose that $f(x) = x^2 - 4x + 5$. What is the ordinate of a point whose abscissa is 3? Wherever x is found in the expression substitute the particular value 3. Then $f(3) = 3^2 - 4 \cdot 3 + 5$. Hence the ordinate $f(3)$ is 2, the result obtained by simplifying the expression. Additional illustrations of this kind should be given by the pupils.

Increment Illustration. It has already been shown that if some point P on the curve represented by y or $f(x) = x^2 - 4x + 5$ has the coördinates (x, y), then $f(x)$ represents the value of the ordinate PM, as shown in Fig. 11. Suppose that the abscissa x is given an increment Δx, then this new abscissa is $x + \Delta x$. Hence wherever x is found in the expression $f(x) = x^2 - 4x + 5$, substitute this new value $x + \Delta x$. Then the expression becomes $f(x + \Delta x) = (x + \Delta x)^2 - 4(x + \Delta x) + 5$, and $f(x + \Delta x)$ is the new ordinate. If this new point is designated by P', then its coördinates are $(x + \Delta x, y + \Delta y)$. This is an important relation because of the way it is used in the explanation of the derivative.

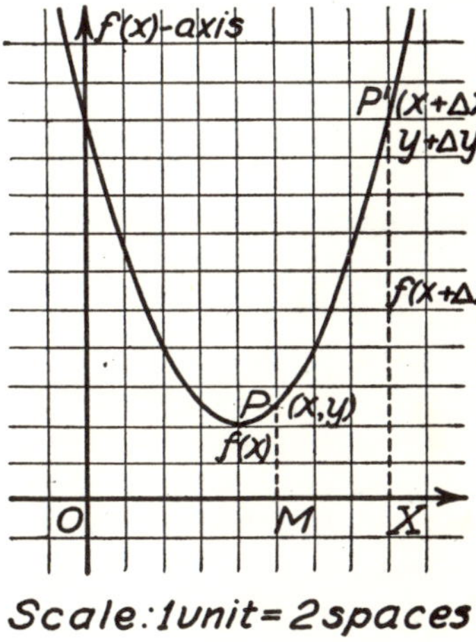

Fig. 11

Examples. Suppose that a railroad company charges n cents per mile traveled. Then the cost of a fare depends upon the distance traveled. Represent the cost by c and the distance by d.

Then $c = f(d)$, that is, the cost is a function of the distance. Now if the distance is increased slightly. how should the expression $c = f(d)$ be altered? It should be written $c + \Delta c = f(d + \Delta d)$. The pupils should give additional illustrations of this kind.

Suppose that $f(x) = x^2 + x + 1$. What is $f(z)$? What is $f(a + b)$? $f(z) = z^2 + z + 1$. $f(a + b) = (a + b)^2 + (a + b) + 1$.

Other Ways of Representing the Function Besides $f(x)$. Suppose that in the same problem, there are two different expressions involving x. One of these may be represented by $f(x)$ and the other by $g(x)$ or some other letter before the parentheses. To distinguish these expressions in talking about them they are read "the f function of x" and "the g function of x."

Example. In solving a practical pioblem these two different functions of x were involved,

$$\frac{100{,}000}{3(x + 1)} + 55x \quad \text{and} \quad \frac{ne(rx^2 + nR) - 2rx \cdot nex}{(rx^2 + nR)^2}.$$

How may these two functions be indicated briefly and yet be distinguishable? This may be done by representing the first one by $f(x)$ and the second by $h(x)$.

Frequently a simple expression is represented by some other form than $f(x)$.

Example. Suppose that $g(x) = x^3 - 1$. What is $g(3)$? $g(3) = 3^3 - 1 = 26$. It should be noted that if a function as $x^3 - 1$ is represented as the g function of x, then the g before (x) in the symbol $g(x)$ must not be changed. Thus, if $g(x) = x^3 - 1$, then $k(x)$, or the k function of x, is a function different from $x^3 - 1$.

Suppose that $s(t) = t^2 - 4t + 5$. Find $s(0)$, $s(1)$, $s(2)$, $s(3)$, and $s(4)$. These values are 5, 2, 1, 2, 5, respectively.

Suppose that $f(y) = y^2 - c^2$. Find $f(1-c)$. *Ans.* $1 - 2c$.

Suppose that $f(B) = 4B + 3$. Find $f(A)$. *Ans.* $4A + 3$.

Suppose that $f(z) = \sin z - \cos z$. Find $f(x)$. *Ans.* $\sin x - \cos x$.

Suppose that $f(u) = \tan u$. Find $f(0.5236)$. *Ans.* $\tan 0.5236$.

Suppose that $f(\#) = \#^3 + 1$. Find $f(5)$. *Ans.* 126.

Suppose that $f(w) = w^5 + w^4 + w^3 + w^2 + 2w + 17$. *Ans.* 23.

Suppose that $f(x) = x^4 - 5x^2$. Find $f(3)$. *Ans.* 26.

The pupils should give many additional examples of this kind in which the function is represented by other letters preceding the parentheses in the symbol, as $m(x)$, $F(z)$, and so on.

THE DERIVATIVE

The derivative is the most important topic in this part of the elementary calculus. The explanation of the derivative as given here is based upon the ideas that were developed in the preceding topics, and for this reason those topics were developed slowly and carefully.

In the explanation of the derivative, the language throughout has been very carefully chosen in order that the mathematical course may stand out clearly. The figure used in the explanation should have on it the fewest possible lines and these should show up all the necessary parts very distinctly. The first step in the meaning of the derivative will now be given.

The First Step. The attention of the class is called to two preliminary facts that will later be used in the explanation of the figure.

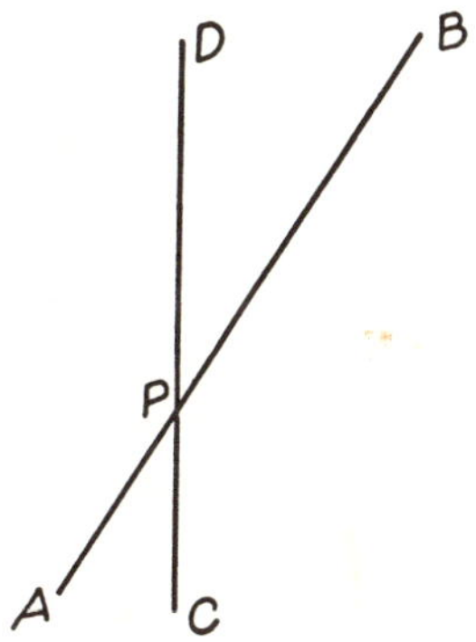

FIG. 12

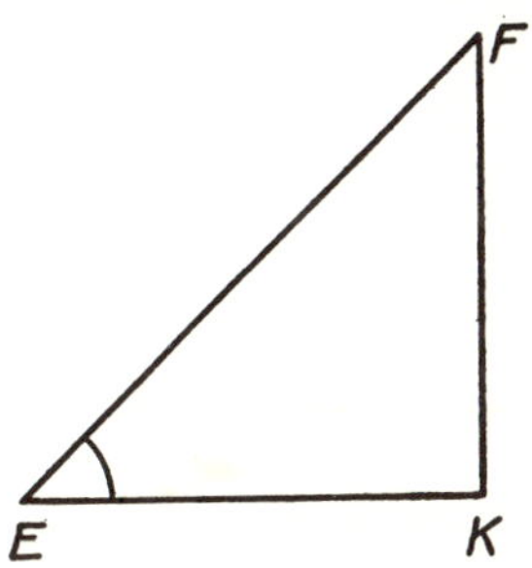

FIG. 13

One of these is that if through any point P on a straight line AB, another line CD is drawn, then CD may be moved about the fixed point P until it coincides with the line AB. The second fact is that in a right triangle EKF the slope of the line EF is the tangent of the angle at E. Also, in the explanation it is well to use the word slope instead of tangent in order to avoid confusing it with the tangent line.

The pupils now plot $f(x) = x^2 - 4x + 5$. Guided by the teacher, they select a suitable point P and through it draw the line PT

tangent to the curve. Next they select a convenient point P', and through P and P' they draw the secant $PP'S$. Then they draw the ordinates PM and $P'M'$, and also PL parallel to the X-axis.

Now consider the right triangle PLQ. It shows that the slope of the tangent line PT is the ratio QL/PL. Now consider the right triangle PLP'. It shows that the slope of the secant $PP'S$ is the ratio $P'L/PL$.

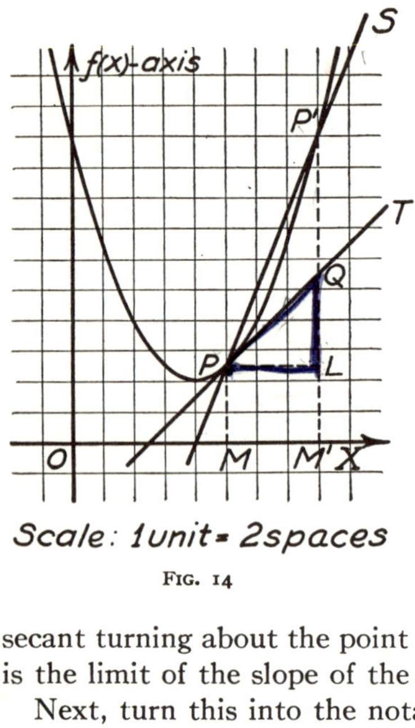

Scale: 1 unit = 2 spaces

FIG. 14

Now imagine that the point P' moves along the curve toward P, then the secant $PP'S$ will more nearly take the position of the tangent line PT. As the point P' gets nearer and nearer to the point P, the position of the secant line will approach yet more nearly to the position of the tangent line, and the slope of the secant line will approach yet more nearly to the slope of the tangent line. In fact, under these conditions the position of the tangent line is the limit of the position of the secant turning about the point P, and the slope of the tangent line is the limit of the slope of the secant.

Next, turn this into the notation of the calculus. Represent the coördinates of P by (x,y). Then by giving x an increment Δx, the coördinates of P' will be $(x + \Delta x, y + \Delta y)$. This idea was developed under the preceding topics. Then, since $PL = \Delta x$ and $P'L = \Delta y$, the slope of the secant is $\Delta y/\Delta x$. Also, as the point P' moves along the curve toward P, Δx approaches zero as its limit. The symbol for expressing the limit of the slope of the secant as Δx approaches zero as its limit is $\lim\limits_{\Delta x \to 0} \frac{\Delta y}{\Delta x}$. This symbol is the geometric ratio form of the derivative. This is read "the limit of the ratio Δy to Δx as Δx approaches zero as its limit."

The Symbol $f'(x)$. At this early stage in the learner's knowledge of the calculus, it would be well to use only $f'(x)$ to represent the derivative of the function of x.[9] This form is used to a large extent in the more advanced calculus and hence it is desirable from the point of view of increased knowledge of mathematical language. Then the derivative may be expressed as $f'(x) = \lim_{x \to 0} \frac{\Delta y}{\Delta x}$. This symbol is a quick way of defining the derivative, gives a desirable point of view, and is not an entirely new way of presenting a definition since the definitions of the trigonometric functions were also explained by stating that they mean such and such things merely because it is agreed that they shall mean those things.

The Second Step. This second step in the meaning of the derivative is the development of Cauchy's fractional form of the derivative. The same curve is used as in the preceding step. The ordinate of P, the point of tangency, PM equals $f(x)$. The ordinate of the moving point P' equals $f(x + \Delta x)$. Now $f(x)$ represents the value of the original function and $f(x + \Delta x)$ represents the value of the function after x has been given an increase. Then the difference between these two ordinates is the increment of the entire function. This difference is $f(x + \Delta x) - f(x)$ and represents the increment of the entire function also. Hence $\Delta y = f(x + \Delta x) - f(x)$, since under the sub-topic, The First Step, it was shown that Δy also represents the increment of the entire function. Substitute this new form of Δy in the definition of the derivative as represented by $\lim_{\Delta x \to 0} \frac{\Delta y}{\Delta x}$.

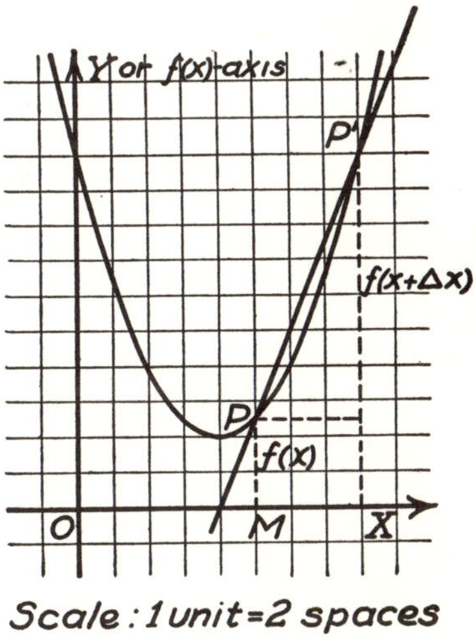

FIG. 15

[9] Throughout the synopsis, the notation has been selected that will most nearly fulfill the requirements of the principle of no exception. The use of the form $\frac{dy}{dx}$ here would be a flagrant transgression of this principle, but after the explanation of the differential, the form $\frac{dy}{dx}$ is then permissible since it may then be regarded as a fraction.

This gives Cauchy's fractional form of the derivative, namely,

$$f'(x) = \lim_{\Delta x \to 0} \frac{f(x + \Delta x) - f(x)}{\Delta x}.$$

The Fundamental Rule. The steps of the fundamental rule for finding the derivative are obtained directly from an analysis of Cauchy's fractional form of the derivative. $f(x + \Delta x)$ stands for the function after x has been given an increment. $f(x)$ is the original function. The difference between these two gives the increment that the entire function has assumed. This difference is divided by the increment of the independent variable. The limit of this resulting fraction is the derivative of the function.

Hence the fundamental rule for finding the derivative is:

a. Find the expression after x has been given an increment.

b. Subtract the original function from this expression.

c. Divide the resulting difference by the increment of the independent variable.

d. Take the limit of this quotient as the increment of the independent variable approaches zero as its limit.

The limit of this quotient is the desired derivative.

Application of the Fundamental Rule. Suppose that $f(x) = 3x + 2$. Find the derivative of $f(x)$.

Give x an increment.	The expression then becomes
a. $3(x + \Delta x) + 2.$	This simplified gives
$3x + 3 \cdot \Delta x + 2.$	From this subtract the original
function $3x + 2.$	This gives
b. $3 \cdot \Delta x$	Divide this expression by Δx. This gives
c. $3 \cdot \Delta x / \Delta x = 3.$	

d. Take the limit of this quotient as Δx approaches zero as its limit. Since 3 is a constant, it is not affected by changes in x. Hence the limit is 3. Therefore the derivative of $3x + 2$ with respect to x is 3. This may be represented by $f'(x) = 3$.

Examples. By means of the fundamental rule, find the derivative of the following:

$f(r) = r^2 + 3.$ *Ans.* $f'(r) = 2r.$

$f(z) = 3z^2 + z.$ *Ans.* $f'(z) = 6z + 1.$

$f(y) = y^3 + 1.$ *Ans.* $f'(y) = 3y^2.$

The pupils should solve additional simple examples. Additional applications of the fundamental rule for finding the derivative will be made in the problems on the slope of the tangent line and in the derivation of formulas.

Remarks. It is hoped that the unity, simplicity, and definiteness in the development of the course thus far is evident. The pupils realize that the ideas of function, increment, independent and dependent variables, function notation, and derivative are not unrelated concepts. The pupils are not lost in the mathematical thought being developed by a discussion on side-topics of theory that are beyond their comprehension, and they are not tired out mentally by a flood of detail given merely to make the discussion exhaustive. Especially is this improvement evident in the development of the idea of the derivative by the use of the three distinct steps: the derivative expressed as the limit of a geometric ratio, $\Delta y / \Delta x$; its interpretation into Cauchy's fractional form of the derivative; and the analysis of this fractional form into the four steps of the fundamental rule for finding the derivative.

It is desirable that the pupils should have a sufficient amount of exercise in finding the derivative of functions by means of the fundamental rule in order that the meaning of the derivative may be impressed upon their minds. After they once begin to use formulas they make very little use of the fundamental rule, but this rule is necessary in order to derive the formulas. Problems on the slope of the tangent form suitable material for exercises in the application of the fundamental rule. Other suitable problems are found in any standard college textbook on the calculus.[10]

THE SLOPE OF THE TANGENT LINE

In the discussion of the first step under the derivative it was shown that the slope of the tangent line is the limit of the slope of the secant, and also that the limit of the slope of the secant is one way of defining the derivative. Hence the value of the derivative of a function at a given point on the curve is the slope of the tangent line at that point.

[10] W. A. Granville, *Differential and Integral Calculus*, p. 31, Boston, 1911.

Plot the graph of $f(x) = x^2 - 4x + 5$. Find the slope of the tangent line at the point $P \equiv (2.5, 1.25)$. To do this, first find the derivative of $f(x)$ by means of the fundamental rule. This gives $f'(x) = 2x - 4$. The value of this derivative at the point P is the slope of the tangent line at this point. The value of the derivative at the point is found by substituting into it the coördinates of the point. Thus, $[2x - 4]^{x=2.5}_{y=1.25} = 2(2.5) - 4 = 1$.

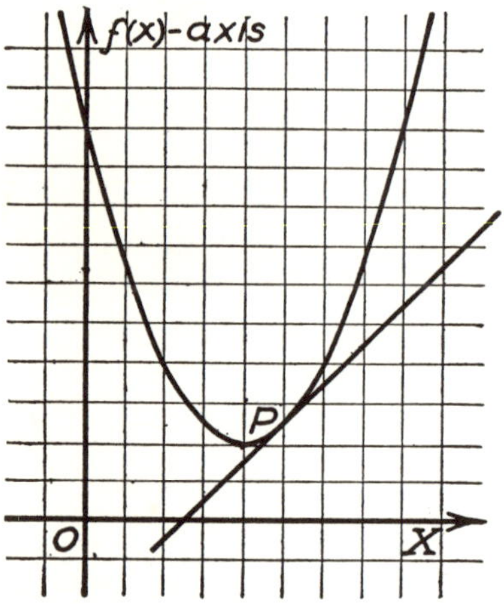

Scale: 1 unit = 2 spaces

FIG. 16

The value of y does not need to be considered here since y does not occur in the derivative. Hence the slope of the tangent line at the point $P \equiv (2.5, 1.25)$ is 1. Since the tan 45° is 1, the angle which the tangent line makes with X-axis is 45°.

Example. Plot the graph of $f(x) = x^2 - 4x + 5$. Find $f'(x)$, the slope of the tangent line at the point whose abscissa is $x = 4$, and also approximately the number of degrees made by the tangent line and the X-axis. Plot the curve.

Ans. (*a*) $2x - 4$; (*b*) its slope is the ratio 4:1; (*c*) the angle is 75° 57.8′.

Suppose that $f(z) = -z^2 - 4z + 5$. Find (*a*) $f'(z)$; (*b*) the slope of the tangent line at the point whose abscissa is $x = -4$; (*c*) approximately the number of degrees in the angle. Plot the curve.

Ans. (*a*) $-2z - 4$; (*b*) its slope is the ratio 4:1; (*c*) the angle is 75° 57.8′.

Suppose that $f(s) = s^2 - 4s + 3$. Find (*a*) $f'(s)$; (*b*) the slope of the tangent at the point whose abscissa is $x = 4$; (*c*) approximately the number of degrees in the angle. Plot the curve.

Ans. (*a*) $2s - 4$; (*b*) its slope is the ratio 4:1; (*c*) the angle is 75° 57.8′.

The pupils should work a sufficient number of similar examples.

FORMULAS

The Differentiation of the Function ku with Respect to x. The process of finding the derivative of a function is called differentiation. The formulas for this purpose are found by means of the application of the fundamental rule. This method will now be illustrated by finding the derivative of ku with respect to x.

Give the variable u an increment. This gives

(1) $k\,(u + \Delta u)$. This simplified gives
$ku + k\cdot\Delta u$. From this subtract the original function
ku. This gives

(2) $k\cdot\Delta u$. Divide this expression by Δx. This gives

(3) $k\cdot\Delta u/\Delta x$. Take the limit of this quotient as Δx approaches zero as its limit. This gives

(4) $\lim\limits_{\Delta x \to 0} k.\frac{\Delta u}{\Delta x}$. Since k is a constant, it is not affected by changes in x. Hence k may be written before the symbol for the limit. Then the expression becomes $k\cdot \lim\limits_{\Delta x \to 0} \frac{\Delta u}{\Delta x}$. The latter part of this expression is the derivative of u with respect to x. Hence the derivative of a constant times a variable is equal to the constant times the derivative of the variable.

Examples. If $f\,(t) = kt$, find the derivative of kt with respect to x by means of the fundamental rule.

The pupils should derive as many additional formulas as the teacher may decide. It is, however, more important that the pupils should learn how to use the formulas.

Examples. Make use of the formulas in solving the following.[11]

Find the derivative of $5x^3$; $4t^2 + 3t + 1$; $av^2 + bv + c$; $7z^4 + 4z^3 + z$; $4q + 3$; $4x^2 + 3x + 2$; $2s^3 + 2s^2 + 3$; $aw^2 + bw + 2$; u^{17}; $ht^a + 3$.

Applications of the Derivative. There is a wide use made of the derivative in practical problems, and illustrations of these may be found in almost any standard college textbook on the calculus.[12] The use of the derivative for finding the slope of the tangent line to a curve has already been shown in this course. Its use will now be shown in the subject of maxima and minima.

[11] Use list of formulas as found in any standard college textbook on the calculus.

[12] William F. Osgood, *Elementary Calculus*, Chap. III, New York, 1921.

MAXIMA AND MINIMA

Maxima and minima form not only a suitable subject for applications of the derivative but also a valuable one from the standpoint of later work in mechanics and other applied subjects. However, the explanations of maxima and minima here will be given by means of graphs in keeping with the idea that the subject-matter of the theory of this course should be closely related and of the same kind throughout.

A Minimum in a Graph. Plot the graph of $f(x) = x^2 - 4x + 5$. Its lowest point is found by inspection. It is the point $P \equiv (2, 1)$. Draw a tangent to the curve at this point. It is seen that this tangent line is parallel to the X-axis. Hence the slope of the tangent line at a point that gives a minimum value of the function is zero. Also, the value of the derivative of the function at this point is zero, since the derivative at a point equals the slope of the tangent line at that point. Now reverse this process. Find the derivative of the function $f(x) = x^2 - 4x + 5$. The derivative $f'(x)$ is $2x - 4$. Set this derivative equal to zero and solve for x. $2x - 4 = 0$. $x = 2$. This is the abscissa of the point P. Find the ordinate corresponding to this abscissa. Since $f(x) = x^2 - 4x + 5$, then $f(2) = 2^2 - 4 \cdot 2 + 5 = 1$. This is the ordinate of the point P. Apparently this is a method that will give the coördinates of the point where the function has a minimum value. It will be shown in the next topic that this same method is used for finding the point where the function has apparently a maximum value. Hence it is necessary to find the way to tell whether the function is a minimum or a maximum at the point whose abscissa is the value of x found by this method. The way to tell the difference will now be shown.

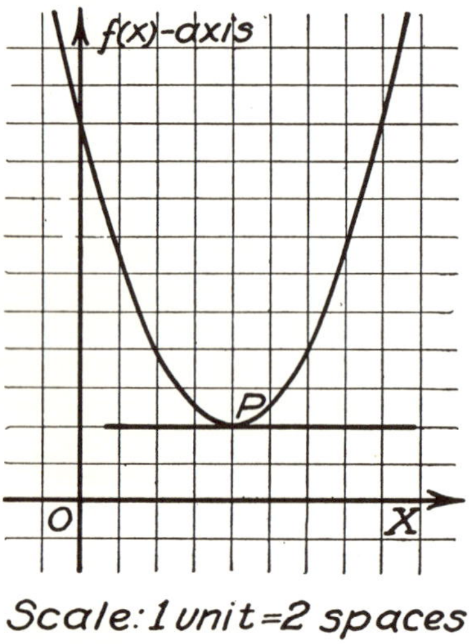

FIG. 17

In the graph of $f(x) = x^2 - 4x + 5$, the point where the function is a minimum is $P \equiv (2, 1)$. This means that the ordinate at the point P is less in value than any other ordinate very close to it. Both positive and negative numbers must be considered. Now use this check on the ordinate of P and see if it really is a minimum value of the function. At P, $x = 2$. This is called a "sign-post" value of x since it shows the way to a minimum or a maximum value of the function. $f(2) = 1$. Suppose that $x = 1.9$, then $f(1.9) = (1.9)^2 - 4(1.9) + 5 = 3.61 - 7.6 + 5 = 1.01$, a value a little greater than 1. It will be found that $f(2.1)$ will also give a value a little greater than 1. Hence 1 or the ordinate of P is a minimum value of the function.

A Minimum Illustrated by a Practical Problem. In practice the nature of the problem will often indicate whether the sign-post value of x leads to a minimum or a maximum value of the function, as is shown in the following printer's problem.

Since the cost of a certain kind of job in printing varied from \$100 to \$200, depending upon the number of electrotypes used, the printer wished to know the most economical number. The conditions of the problem are as follows: 200,000 (P) prints are required; 1200 (S) prints per hour is the speed of the press; \$2.00 (R) per hour is the cost of running the press; 55 cents (E) each is the cost of the extra electrotypes needed after the type is once set up. Required the number of electrotypes (x) that should be used to secure the minimum cost (C). The problem is evidently one in determining the minimum value of C by the use of the differential calculus. Here C is a $f(x)$. Solution:

$f(x) = \dfrac{200{,}000 \cdot 200}{1200\,(1 + x)} + 55x = \dfrac{100{,}000}{3\,(x + 1)} + 55x$, where x is the number of electrotypes. By differentiation, $f'(x) = -100{,}000/3$ times $\dfrac{1}{(x + 1)^2} + 55 = 0$. This derivative is set equal to zero in order to find the sign-post value of x. Hence $x = \dfrac{100}{3} \cdot \sqrt{66} - 1 =$ 23.6 +. This value of x will make the function a minimum. There is no maximum value of the function since it is seen that as x becomes larger and larger the value of the function becomes larger and larger.

Therefore the most economical number of electrotypes to use for this purpose is 24.[13]

Examples. A Labrador missionary-doctor is in his rowboat 4 miles from shore. A messenger informs him that a man 6 miles down the coast is in extreme need of his services. The doctor can row at the rate of 3 miles per hour. On shore, the doctor can make an average speed of 4 miles an hour on his skiis. How far down the shore should he land in order that he may reach his patient in a minimum length of time? The distance will be found to be 4½ miles.

An architect wishes to design a two-story type-cottage with pitched roof such that the cross-section of the room space on the second floor will be 12 ft. wide and 6 ft. high. What will be the minimum length of the rafters? Ans. 16.968 ft. or 17 ft. Suggestion. Consider the right triangle formed by a rafter, the perpendicular distance from the ridge to the second floor, and half the width of the cottage. The cross-section of half the room space will be a square inscribed in this right triangle with a side 6 ft. in length. Then use the proportions resulting from the properties of similar triangles.

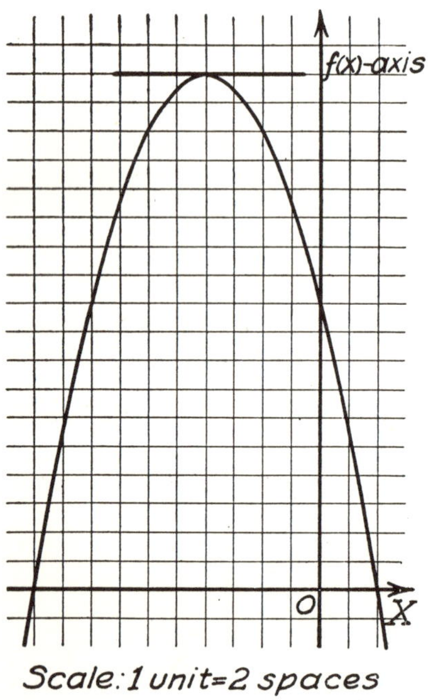

FIG. 18

A Maximum in a Graph. Plot the graph of $f(x) = -x^2 - 4x + 5$. Its highest point is found by inspection of the figure. It is the point $P \equiv (-2, 9)$. Draw a tangent to the curve at this point. It is seen that this tangent line is parallel to the X-axis, and hence the slope of the tangent line at a point that gives a maximum value of the function is zero. Also, the value of the derivative of the function at this point is zero, since the derivative

[13] Edgar E. DeCou, "A Practical Printer's Problem", *American Mathematical Monthly*, V. 27, Nov. 11, 1920.

at a point equals the slope of the tangent at that point. Now reverse the process. Find the derivative of the function $f(x) = -x^2 - 4x + 5$. The derivative $f'(x) = -2x - 4$. Set this derivative equal to zero and solve for x.

$-2x - 4 = 0$. $x = -2$. This is the abscissa of the point P. Find the ordinate corresponding to this abscissa. Since $f(x) = -x^2 - 4x + 5$, then $f(-2) = -(2)^2 - 4(-2) + 5 = 9$. This is the ordinate of the point P as found by inspection. Apparently this is a method that will give the coördinates of the point where the function has a maximum value. It has, however, already been shown that this same method is used for finding the point where the function has a minimum value. It was also shown how to use the sign-post value of x to tell if it will give a minimum value of the function. How to tell if it will give a maximum value of the function will now be shown.

In the graph of $f(x) = -x^2 - 4x + 5$, the point where the function is a maximum is $P \equiv (-2, 9)$. This means that the ordinate at the point P is greater in value than any other ordinate very close to it. Both positive and negative numbers must be considered. Now use this check on the ordinate of P and see if it really is a maximum value of the function. At P, $x = -2$, and the ordinate $f(-2)$ is 9. Find the ordinates $f(-2.1)$ and $f(-1.9)$. It is seen that each of these is less than the ordinate $f(-2)$. Hence P is a point where the function has a maximum value.

A Maximum Illustrated by a Practical Problem. Suppose that a tinsmith has on hand ends of strips of tin left over from some job. He wishes to make a box with each piece for holding such things as nails of different sizes. Each piece is 15 inches by 8 inches. The boxes shall be rectangular in shape and shall contain the greatest volume possible. They are made by cutting out a square piece from each corner and then turning up the sides. What should be the size of the piece cut out of each corner?

First we must form the expression of which we wish to find the maximum. Let x = the size of the square to be cut out of each corner. Then $15 - 2x$ = the length of one side of the box, and $8 - 2x$ = the width of the box, and x = the depth of the box.

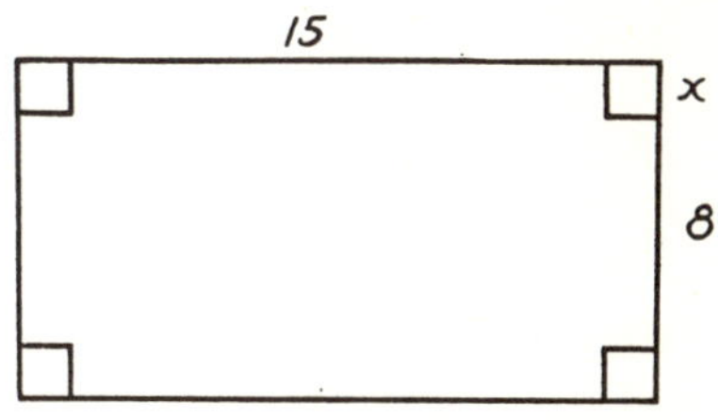

FIG. 19

Then the volume is $x(8 - 2x)(15 - 2x)$. This is the expression of which we are to find a value of x so that the volume will be a maximum. Expanding this expression, we have $f(x) = 4x^3 - 46x^2 + 120x$. Its derivative is $f'(x) = 12x^2 - 92x + 120$. Set the derivative equal to zero and solve for x.

$$12x^2 - 92x + 120 = 0,$$
$$3x^2 - 23x + 30 = 0.$$
$$x = (23 \pm \sqrt{529 - 360})/6 = 6, \text{ or } 1\tfrac{2}{3}.$$

We see that $x = 6$ inches is not possible. Hence we select $x = 1\frac{2}{3}$ inches. Then if the box is constructed according to these conditions, its volume will be the maximum. Check this result by plotting $f(x) = x(8 - 2x)(15 - 2x)$. This factor form makes it easier to substitute different values for x for plotting.

Examples. Given the conditions the same as in the preceding problem except that the length is 16 inches and the width is 6 inches. Find the value of x that will make the volume a maximum. It will be found that this value of x is one and one-third inches.

A farmer wishes to build a concrete cylindrical cistern that shall contain 5,000 cu. ft. of water. What should its dimensions be in order that the cost of the material will be a minimum? The surface is made up of the lateral surface and the base of the cylinder. It will be found that the radius should be made equal to the height or 11.7 ft.

Maxima and Minima in Graphs. A graph may show a number of points where the function has a maximum value or where it has a minimum value. An example illustrating this will now be given.

Find the points where the function $f(x) = x^3 - 2x^2 - 5x + 6$ has either a maximum value or a minimum value. Its derivative is $f'(x) = 3x^2 - 4x - 5$. Set this equal to zero and solve for x. Then $x = 2.1$, or -0.8. Here there are two sign-post values of x. First consider the value -0.8. If the abscissa x is -0.8, then the ordinate $f(-0.8)$ is $(-0.8)^3 - 2(-0.8)^2 - 5(-0.8) + 6$, or 8.2. That this is a maximum value of the function may be seen by finding ordinates corresponding to values of the abscissa x that are a little greater and then a little less than -0.8. It will be found that these ordinates are less than the ordinate $f(-0.8)$. Hence $f(-0.8)$ is a maximum value of the function.

If the abscissa is $x = 2.1$, then the corresponding ordinate $f(2.1)$ is a minimum value of the function. Verify this by the usual checking method. Hence $f(x) = x^3 - 2x^2 - 5x + 6$ has a maximum value $f(-0.8)$ for the value of $x = -0.8$, and a minimum value $f(2.1)$ for the value of $x = 2.1$.

Suggestions for Plotting the Curve

The important points in plotting a graph are the place where it crosses the $f(x)$-axis, the place where it has a maximum or a minimum value, and the place where it crosses the X-axis. An illustration of how to do this will now be given.

Plot the graph of $f(x) = x^3 - 2x^2 - 5x + 6$.

At the place where the curve crosses the $f(x)$-axis, the abscissa $x = 0$. Since $f(0) = 0^3 - 2(0)^2 - 5(0) + 6 = 6$, this is the place where the graph crosses the $f(x)$-axis.

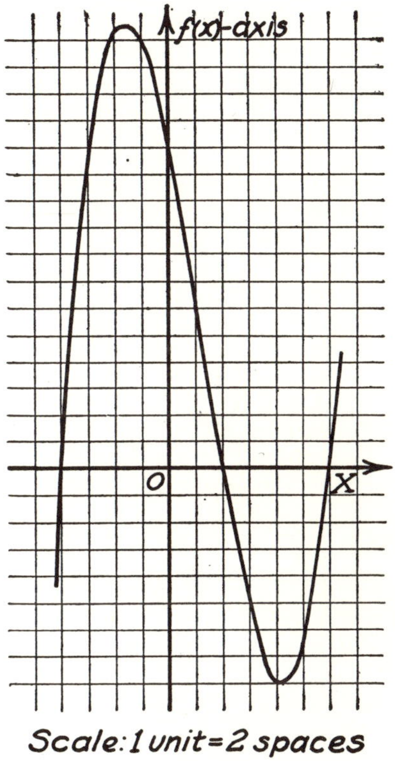

FIG. 20

The function has a maximum value at the point $(-0.8, 8.2)$ and a minimum value at the point $(2.1, -4)$.

The places where the function crosses the X-axis are the roots of the equation. These are $x = 1$, $x = 3$, and $x = -2$.

The general shape of the graph may now be drawn, as shown in Fig. 20. Additional points may be found if a more careful graph is desired.

Example. Find the values of x for which the following function will be a maximum and for which it will be a minimum. The function is $f(x) = x^3 - 6x^2 + 12$. It will be found that the function has a maximum value when $x = 0$, and a minimum value when $x = 4$.

THE DIFFERENTIAL

Reasons for Studying the Differential. By means of the differential we pass from differentiation to integration and reverse the process. A knowledge of the differential also enables pupils to extend the skill they have thus far acquired to other problems. The idea of the differential is made simpler by explaining it in such a way that pupils are able to visualize it.

Geometrical Representation. Plot the graph of $y = x^2 - 4x + 5$. The slope of the tangent line is QL/PL, and the slope of the secant is $P'L/PL$. These are used here under the same conditions under which they were used in the explanation of the derivative. The coördinates of P are (x, y) and of P' are $(x + \Delta x, y + \Delta y)$. Then $P'L$ represents Δy. It has been shown that the limit of the slope of the secant is the slope of the tangent line, as Δx approaches zero as its limit. Now if the point P' is taken in some particular position on the curve, as shown in the accompanying illustration, then Δy is made up of two parts, $P'Q$ and QL. Only the latter, QL, will be considered here. The increment of the entire function is $P'L = \Delta y$. The part QL of this increment is called the differential of the function. It is represented by the symbol, dy, and is read "the differential of y." The increment of the independent variable $PL = \Delta x$. This is equal to dx, as is shown in the more advanced work in the calculus. However, it is seen from the explanation here that dy does not equal Δy. It will be shown in the next sub-topic what dy really does equal. Since these quantities, dy and dx, are finite, they may be substituted for their equals in the fraction, QL/PL, which represents the slope of the tangent line. Hence the slope of the tangent line at a particular point is dy/dx.

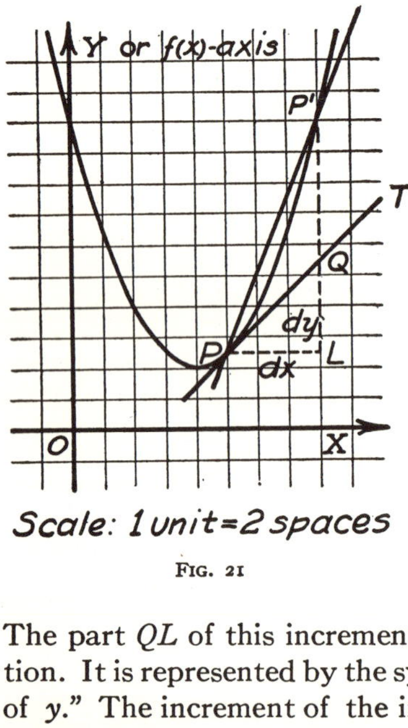

Scale: 1 unit=2 spaces

FIG. 21

What dy Equals. It has previously been shown that the slope of the tangent line at a particular point of the curve is the derivative of the function or the value of $f'(x)$ at that particular point. In the preceding sub-topic it was shown that the slope of the tangent line is the value of dy/dx at that point of tangency. Hence the value of dy may be found by setting these two quantities equal to each other and solving the equation for dy. Thus,

$$dy/dx = f'(x), \quad \text{then} \quad dy = f'(x) \cdot dx.$$

Hence the differential of a function is found by multiplying the derivative of the function by the differential of the independent variable. If this method is adhered to, it will not be necessary to make use of a separate table of differentials.[14]

Examples. Find the differentials of the following functions.

$f(x) = 5x^2$. Then $df(x) = 10xdx$.
$f(t) = 4t^3 - 3t^2 - 1$. Then $df(t) = 12t^2dt - 6tdt$.

Find the differentials of the following:

$5x^2$,	*Ans.* $10xdx$.
$4t^3 - 3t^2 - 1$,	*Ans.* $12t^2dt - 6tdt$.
$av^2 - b$,	*Ans.* $2avdv$.
$7z^3 - z$,	*Ans.* $21z^2dz - dz$.
kt,	*Ans.* kdt.
$\frac{1}{2}gt$,	*Ans.* $\frac{1}{2}gdt$.
$ht^a - 2t - 3$,	*Ans.* $aht^{a-1}dt - 2dt$.
cx^{n-1},	*Ans.* $(n-1)cx^{n-2}$.

The pupils should solve many additional examples of this kind.

INTEGRATION

A Reason for Studying Integration. One of the reasons for studying integration is that it enables one to find the areas of figures which cannot be found by the ordinary geometric methods. Plot the curve $f(x) = x^2 - 4x + 5$. Erect ordinates at $x = a$ and $x = b$. How can the area enclosed by these ordinates, the curve, and X-axis be found? This will now be shown.

[14] The general treatment of the differential of which the method here given is an original adaptation is found in Granville's *Calculus*, p. 132.

At the point P the abscissa x equals a. Give x a small increment. Represent this by dx. This may be done since under the topic of the differential it was pointed out that the increment of the independent variable x equals dx. At the right-hand extremity of dx erect an ordinate. Then this small strip of the area is almost rectangular in shape with a base equal to dx and an altitude approximately equal to the ordinate $f(x)$. Then the area of this strip is approxiately $f(x) \cdot dx$. Next divide the remainder of the area into strips each with its base equal to dx. Let x assume these different points of division considered as abscissas along the X-axis. Then for each of these values of x the corresponding values of the ordinates will be $f(x)$, and the area of each of these strips will be represented by $f(x) \cdot dx$. The sum of all such strips is approximately the entire area desired. In fact, it is proved in the more advanced calculus that the limit of this sum as dx becomes infinitesimally small is the exact area of the figure. This limit of the sum of these strips is represented by $\int f(x)dx$. The mark $\int$ is a modification of the letter S, the first letter of the word sum. The symbol is read "the integral of $f(x)dx$." The word integral means the whole and as used here it represents the sum of all the parts. In case there are definite limits to the value of x as in the illustration, the symbol is written $\int_a^b f(x)dx$, where a stands for the lower limit of x and b stands for the upper limit of x.

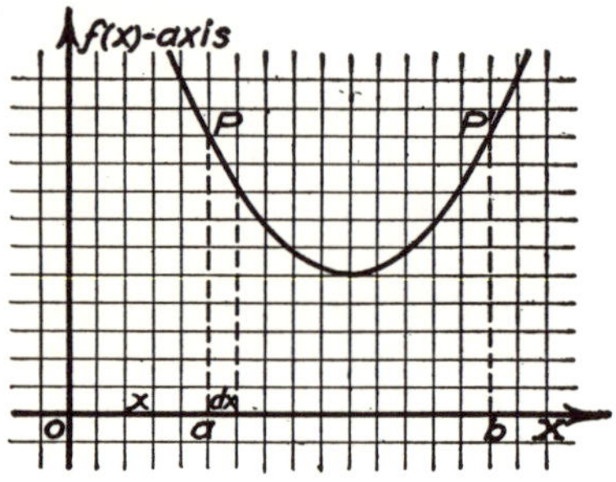

Scale: 1 unit = 5 spaces

FIG. 22

Integration the Reverse of Differentiation. Suppose that we consider only the direction along the positive end of the X-axis. Take n points of division on the line beginning with zero and with the distance between any two successive points equal to dx. Then the value of x_n would be the integral of dx, that is, the sum of all these elementary divisions. Now let x represent in general any point along the X-axis, then the value of x may be represented by $\int dx$.

FIG. 23

This may be written $x = \int dx$. Apparently the quantity under the integral sign is the differential of the expression to which the integral is equal. In fact, it is true that the value of the integral is the expression which when differentiated will give the quantity under the integral sign. Thus, it is seen that integration and differentiation are reverse processes, somewhat like division and multiplication, since one may be obtained from the other.[15] The method for finding the integral when the differential is given will now be taken up.

A First Attempt at a Table of Integrals. First put down some simple graded functions and their differentials, as

dx is the differential of x,
xdx is the differential of $\frac{1}{2}x^2$, and so on.

Since the function from which we obtained the differential is the integral of that differential, the table may be written,

x is an integral of dx,
$\frac{1}{2}x^2$ is an integral of xdx, and so on.

However, if the function from which the differential was obtained contains a constant term, this term disappears entirely in the differential since the differential of a constant is zero. Hence when we reverse the process for finding the integral, there is no way of knowing what this constant term was in the original function. Thus,

the differential of $x + 5$ is $dx + 0$, and
the differential of $x + 7$ is $dx + 0$. Then the

integral of dx is x plus or minus a constant. Hence in reversing the process of differentiation, the constant term in the integral is indeterminate. Integrals in which the value of the constant term is not known are called indefinite integrals. Sometimes the conditions of the problem are such that there are no indeterminate constant terms in the integral and the integral is taken between definite limits of the independent variable. Such integrals are called definite integrals and will be explained later.

If the differentials and their corresponding integrals are arranged in order, a general formula may be derived. An illustration of this will now be given. (Bernoulli's table.)

[15] This method was suggested by Professor D. E. Smith in discussing an experiment on integration in the Horace Mann School, Teachers College.

The integral of adx is ax $\pm$ a constant,
The integral of $axdx$ is $\frac{1}{2}ax^2$ $\pm$ a constant,
The integral of ax^2dx is $\frac{1}{3}ax^3$ $\pm$ a constant,
The integral of ax^3dx is $\frac{1}{4}ax^4$ $\pm$ a constant, and so on.

Hence the integral of ax^pdx is $\dfrac{a}{p+1} . x^{p+1}$ $\pm$ a constant.

The pupils should be given practice in the use of the integral tables in finding the integrals of given differentials.

Examples. Find the integrals of the following:

$10xdx,$	*Ans.* $5x^2 + c.$
$3t^2dt - 2tdt,$	*Ans.* $t^3 - t^2 + c.$
$6avdv,$	*Ans.* $3av^2 + c.$
$15z^2dz - 2dz,$	*Ans.* $5z^3 - 2z + c.$
$at^{a-1}dt - 2tdt,$	*Ans.* $t^a - t^2 + c.$

THE DEFINITE INTEGRAL

In the beginning of the discussion on integration it was shown that $\int_a^b f(x)dx$ represented the area bounded by the curve, the X-axis, and the ordinates erected at a and b on the X-axis. Then the range of values that x assumes is the segment of the X-axis from a to b, or $b - a$. To find the value of the integral first substitute b for x in the integral and then a, the difference between these two results taken in this order is the value of the integral. This too is the value of the area which was to be found. The constant of integration does not need to be considered in the definite integral since it cancels out.

Example. In the illustration just used, suppose that $a = 1$, $b = 3$, and $f(x) = x^2 - 4x + 5$. Use Fig. 22.

Solution. Substitute the given values in the formula for the area which is $\int_a^b f(x)dx$. This gives

$$\int_1^3 (x^2 - 4x + 5) = \int_1^3 x^2dx - \int_1^3 4xdx + \int_1^3 5dx = \left[\frac{x^3}{3} - \frac{4}{2}x^2 + 5x\right]_1^3.$$

This last form shows how to indicate the limits that are to be substituted in the result. The rule for substitution is: "Substitute the upper limit for the variable in the expression in brackets and then subtract from this result the result obtained by substituting the lower limit for the variable in the expression in brackets."

Thus, $(\frac{3^3}{3} - 2 \cdot 3^2 + 5 \cdot 3) - (\frac{1}{3} - 2 + 5) = (9 - 18 + 15) - (\frac{1}{3} - 2 + 5) = 6 - 3\frac{1}{3} = 2\frac{2}{3}$.

Hence the area is $2\frac{2}{3}$ square units. Check this result by means of squared paper.

Examples. Suppose that the conditions are the same as in the preceding problem except that the limits are from the ordinate 2 to to the ordinate 4. Find the area bounded by the curve, the X-axis, and these two ordinates.

Suppose that $f(x) = -x^2 - 4x + 5$. Find the area bounded by this curve and the X-axis between the limits $x = -5$ and $x = 1$.

Suppose that $f(x) = x^3 - 2x^2 - 5x + 6$. Find the area bounded by: (*a*) the curve, the $f(x)$-axis, and the X-axis, the limits being $x = 2$ and $x = 0$; (*b*) the curve, the $f(x)$-axis, and the X-axis, the limits being $x = 0$ and $x = 1$; (*c*) the curve and the X-axis, the limits being $x = 1$ and $x = 3$.

These examples show the use of the definite integral for finding the area of a plane surface. In addition to this application of the definite integral, it may be used for finding the areas of curved surfaces, the lengths of arcs, the volumes of solids, centers of gravity, and moments of inertia. The application of the definite integral for finding the length of an arc and for finding the volume of a solid of revolution will now be taken up.

THE LENGTH OF AN ARC

The Formula. The definite integral can also be used for finding the length of an arc. Suppose that it is desired to find the length of the arc in the accompanying figure from the point on the curve whose abscissa is a to the point whose abscissa is b. Take a small segment of the arc at the point whose abscissa is a, and call this increment of the arc ds. Then this small arc ds is the hypotenuse of the right triangle whose sides are dy and dx. The arc ds is taken so very small that there will be no appreciable error if we treat it as though it were a straight line. Then $ds^2 = dy^2 + dx^2$ since the square on the hypotenuse is equal to the sum of the squares on the other two sides. Under the topic of the differential it was shown that $dy = f'(x)\,dx$. Substitute this form of the differential of the function in $ds^2 = dy^2 + dx^2$. Then the expression becomes

$$\begin{aligned} ds^2 &= [f'(x)\,dx]^2 + dx^2 \\ &= [f'(x)]^2 \cdot dx^2 + dx^2 \\ &= \{[f'(x)]^2 + 1\} \cdot dx^2. \end{aligned}$$

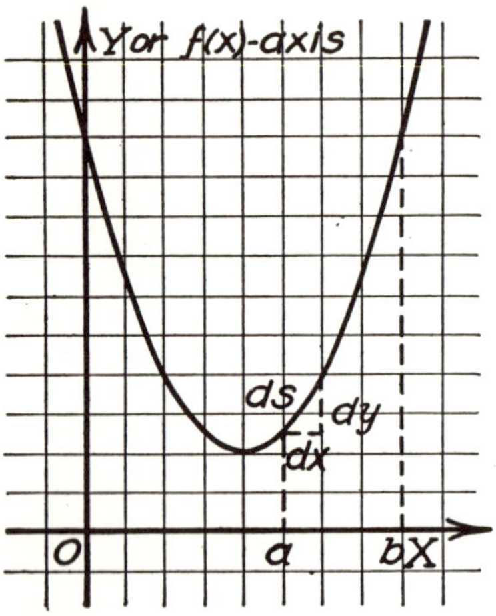

Scale: 1 unit = 2 spaces

FIG. 24

By taking the square roots of both sides of this equation, it becomes $ds = \sqrt{f'(x)^2 + 1} \cdot dx$.

This is only an increment of the arc. Then the length of the entire arc will be the sum of all these increments as x assumes all the values of the range on the X-axis between the limits a and b. This sum is an integral and is represented thus, $s = \int_a^b \sqrt{f'(x)^2 + 1} \cdot dx$

Hence to find the length of an arc between definite limits, substitute the derivative of the function in this formula and then find the integral of the resulting expression.

An Application of the Formula. Find the circumference of the circle $f(x) = \sqrt{(r^2 - x^2)}$.

Divide the circle into four quadrants and find the length of the arc of one of them. Four times this result will give the entire circumference.

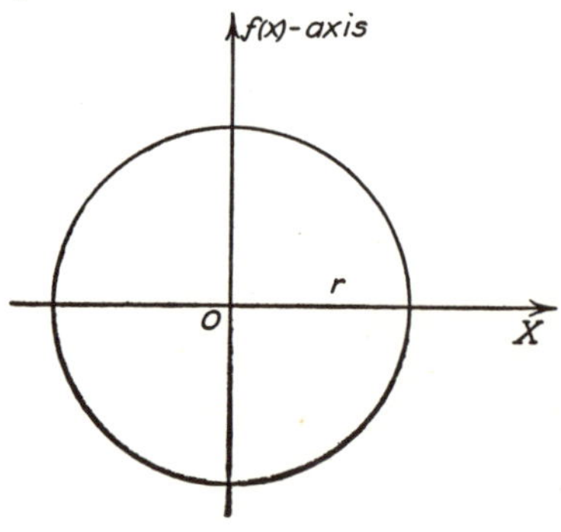

FIG. 25

The limits are $x = 0$ and $x = r$. Hence $a = 0$ and $b = r$. $f'(x) = -x/f(x) = -x/\sqrt{(r^2 - x^2)}$. Then $f'(x)^2 = x^2/(r^2 - x^2)$. By substituting these values in the formula, it becomes

$$s = \int_0^r [x^2/(r^2 - x^2) + 1]^{1/2} \cdot dx$$

$$= \int_0^r \left[\frac{x^2 + r^2 - x^2}{r^2 - x^2}\right]^{1/2} \cdot dx$$

$$= \int_0^r r/(r^2 - x^2)^{1/2} \cdot dx$$

$$= r\,[\text{arc sin } x/r]_0^r$$

$$= r\,(\text{arc sin } 1 - \text{arc sin } 0)$$

$= r\,(\pi/2 - 0) = r \cdot \pi/2$. Then the entire circumference is four times $r \cdot \pi/2$, or $2\,\pi\,r$.

Examples. By integration find the circumference of the following circle: $x^2 + y^2 = 9$. It will be found to be 6 times 3.1416.

Find the length of an arc of the parabola $f(x) = x^2 - 4x + 5$ with the limits $x = 1$ and $x = 3$. Use formula No. 124 in Peirce's *A Short Table of Integrals*, and check by using formulas Nos. 165 and 160.

THE VOLUME OF CERTAIN SOLIDS

The Formula. Suppose that we have given a rectangle $OABC$ in the position as shown in the figure.

Then if we rotate the rectangle about O-X as an axis, the rectangle will generate a cylinder.

Now use dx as a unit of measure and divide OA into as many equal parts as dx will be contained into it.

Then pass planes through each of these points of division.

Then these planes will divide the cylinder into many equal cylinders each with an altitude equal to dx and a base equal to the base of the original cylinder. OC, or the radius of this circle, or base, is equal to $f(x)$. Then the area of this circle is $\pi\,[f(x)]^2$, since the area of a circle is π times the square of the radius. Hence the volume of each cylinder of altitude dx is $\pi \cdot [f(x)]^2 \cdot dx$.

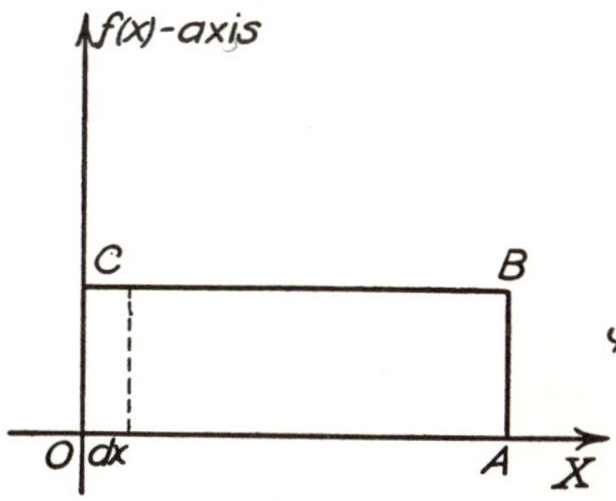

FIG. 26

The sum of all such volumes would approximately equal the volume of the cylinder, and the limit of this sum as dx approaches zero as its limit is the volume. This limit is the integral $\int \pi[f(x)]^2 dx$. Hence, the volume of a solid formed by rotating a plane figure about an axis is $\pi \int_a^b [f(x)]^2 dx$ where the upper and lower limits of x are b and a respectively.

Since the generating line CB is parallel to the X-axis, $f(x)$ is the same no matter what values are given to x and is always equal to OC. Represent OC by a given letter r. Then $f(x) = r$. Represent the length of the cylinder by h, then the limits of x are o and h. Substitute these values in the formula $\pi \int_a^b [f(x)]^2 dx$. This gives $\pi \int_o^h r^2 dx$. The integral of this is $\pi r^2 x]_o^h$. The result after substituting the limits is $\pi \cdot r^2 h$. This expressed in words is the customary rule for finding the volume of a cylinder, namely, its volume is equal to the product of the area of the base by the altitude.

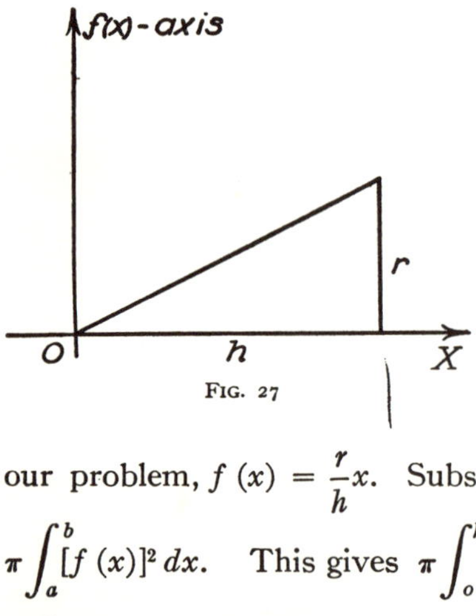

FIG. 27

An Application of the Formula. Find the volume of a cone of radius r and height h.

The generating line has a constant slope. It is r/h, and it intersects the $f(x)$-axis at the origin or zero. Then in the slope form of the straight line, $f(x) = mx + b$, $m = r/h$ and $b = o$. Hence in our problem, $f(x) = \frac{r}{h}x$. Substitute this value in the formula, $\pi \int_a^b [f(x)]^2 dx$. This gives $\pi \int_o^h \frac{r^2}{h^2} \cdot x^2 dx$. The value of this integral is $\frac{1}{3}\pi \cdot r^2 h$. In words, this is the customary rule for finding the volume of a cone, namely, the volume of a cone is equal to the product of the area of the base times one-third of the altitude.

Example. Suppose that we have given a semi-circle with radius r. This semi-circle rotated about its axis will generate a sphere. Find the volume of this sphere. Its volume is $\frac{4}{3} \cdot \pi . r^3$.

GENERAL CONCLUSIONS

From the study as here set forth, it is evident that a first course in the elementary calculus is not too difficult for senior high-school pupils. This conclusion is based largely upon the following facts:

1. A historical survey of the natural growth of the calculus in the development of mathematics shows that the subject has adapted itself slowly to the needs of the race, and that a similar adaptation to the needs of those pupils in the senior high schools who care to elect the subject is entirely possible;

2. An examination of those leading European textbooks on the elementary calculus which have been written for beginners shows that a number of the best teachers of secondary mathematics do not consider the elementary calculus too difficult for their pupils;

3. The theory and the applications of the elementary calculus can be so simplified that it becomes an interesting and extremely valuable subject for such students in the senior high school as show fair ability in mathematics.

It is therefore evident that a course in the elementary calculus should be included in the elective mathematics of the senior high school for the following reasons:

1. It is necessary if our American pupils are to have an equal opportunity in mathematics with pupils in schools abroad;

2. Such a course is needed to satisfy the demand that has been created for additional mathematics in the senior high school by the trend of mathematics in our public school system;

3. The elementary calculus is one of the most important aids in applied mathematics and contributes largely to the field of pure mathematics;

4. Such a course will enable the pupil who intends to enter any field of applied mathematics to continue further his study of the calculus, as found in such a field, without the aid of a teacher.